BONOBO

THE FORGOTTEN APE

BONOBO

THE FORGOTTEN APE

FRANS DE WAAL

PHOTOGRAPHS

FRANS LANTING

UNIVERSITY OF CALIFORNIA PRESS
Berkeley Los Angeles London

University of California Press
Berkeley and Los Angeles, California

University of California Press, Ltd.
London, England

Library of Congress Cataloging-in-Publication Data

Waal, F. B. M. de (Frans B. M.), 1948–
Bonobo : the forgotten ape / Frans de Waal ; photographs, Frans Lanting.
p. cm.
Includes bibliographical references and index.
ISBN 0-520-20535-9 (cloth : alk. paper)
1. Bonobo. 2. Bonobo—Behavior. I. Lanting, Frans. II. Title.
QL737.P96W3 1997
599.88′44—dc20 96-41095
CIP

Printed in Hong Kong
9 8 7 6 5 4 3 2

This book is printed on acid-free paper.

CONTENTS

An inquisitive bonobo. Note the typically narrow shoulders and thin neck, as well as the relatively small, rounded skull, which first suggested to anatomists that this might be a different species. Initially bonobos were known as "pygmy chimpanzees," but most scientists now find this name misleading, given the overlap in weight between bonobos and chimpanzees. More gracile than their cousins, bonobos have reddish lips in a black face and thin black hair on their heads. Captive bonobos spend so much time grooming one another, however, that some of them lose their hair. The baldness is only temporary; this female grew her hair back after being introduced to new companions, who groomed her less often.

OVERLEAF: *Bonobos at the edge of a forest near Wamba. Other populations in the rain forest of Zaire also need to be studied, because bonobos do not necessarily act the same throughout their range. But only at Wamba have data been collected continuously for more than twenty years.*

There are only about one hundred captive bonobos in the world, compared to thousands of captive chimpanzees. Zoo breeding programs have made giant strides towards improved health and reproduction for their precious bonobos, as well as better environments. This indoor enclosure at the Cincinnati Zoo mimics the natural habitat.

To our gracile relatives,

and the people who have devoted their lives

to discovering, studying, and protecting them.

PREFACE

How better to celebrate a little-known close relative than by marrying science and art, combining hard-won information with provocative images? The result is the first book that tries to piece together the bonobo puzzle in a fashion that we hope will appeal to a wide audience.

The odyssey towards the current state of knowledge began long ago, on several continents. Following the discovery of the species in a Belgian museum, in 1929, early studies at German zoos revealed how much the bonobo's behavior differed from that of its sibling species, the chimpanzee. Next came the first expeditions of Japanese and Western scientists to the interior of Zaire to penetrate the social life of this elusive anthropoid. Thanks to sustained efforts in the field and further research in captivity, we are finally catching a first glimpse of the bonobo's unique society. But it is still just a glimpse. Few outsiders have ever seen a live bonobo in the wild, and photographs of them are exceedingly rare.

Recent advances in research prompted the two of us, a zoologist and a photographer, to collaborate on this project. Despite our different professions, our backgrounds are similar. We are both from a generation of Dutch naturalists who grew up at a time when nationally celebrated novelists, poets, and even television personalities were inspired by the ideas of Konrad Lorenz and Niko Tinbergen, who later shared a Nobel Prize for their revolutionary studies of animal behavior. Ethology, as the field has become known, promotes a broad view of animals. They are studied not so much as models for ourselves but as adaptation artists. Each species has evolved its own form of communication, its own way of dealing with the problems posed by its environment, and its own social organization, within which it survives and reproduces.

Bonobo behavior has been studied ethologically from two complementary perspectives. Field research has produced invaluable information about the species' natural history, while observations in enlightened zoo settings have supplemented the necessary details on behavior. So much has been learned from the pathbreaking work of only a dozen experts that the bonobo has become quite a hot topic among primatologists—but unfortunately, theirs is a tiny community. While everyone knows chimpanzees, gorillas, and orangutans from documentaries and magazines, few people have even heard of bonobos. Books and articles on the other apes easily fill a small library; for a complete collection of literature on bonobos, a single cardboard box will do.

It is high time for increased public awareness of this appealing, fascinating primate, one that presents a major challenge to traditional notions of human origins. This challenge coincides with recent fossil finds that are undermining entrenched views about how life on the savanna shaped human evolution. Evidence now suggests that bipedal locomotion, a transition long regarded as the defining moment in human prehistory, coexisted for some time with an arboreal lifestyle—a finding that could make the forest-dwelling bonobos key to new reconstructions of the past.

The bonobo is not a historical phenomenon, however—at least not yet. The species still lives in a region of the world whose remoteness has thus far protected it from the large-scale habitat destruction that is taking place everywhere else in the tropics. If we take good care of the bonobo, we may for a long time share this planet with a family member that affords us an entirely new look at ourselves.

Frans B. M. de Waal
Atlanta, Georgia

Frans Lanting
Santa Cruz, California

CHAPTER 1

THE LAST APE

When the lively, penetrating eyes lock with ours and challenge us to reveal who we are, we know right away that we are not looking at a "mere" animal, but at a creature of considerable intellect with a secure sense of its place in the world. We are meeting a member of the same tailless, flat-chested, long-armed primate family to which we ourselves and only a handful of other species belong. We feel the age-old connection before we can stop to think, as people are wont to do, how different we are.

Bonobos will not let us indulge in this thought for long: in everything they do, they resemble us. A complaining youngster will pout his lips like an unhappy child or stretch out an open hand to beg for food. In the midst of sexual intercourse, a female may squeal with apparent pleasure. And at play, bonobos utter coarse laughs when their partners tickle their bellies or armpits. There is no escape, we are looking at an animal so akin to ourselves that the dividing line is seriously blurred.

Whereas the bonobo amazes and delights many people, the implications of its behavior for theories of human evolution are sometimes inconvenient. These apes fail to fit traditional scenarios, yet they are as close to us as chimpanzees, the species on which much ancestral human behavior has been mod-

Many primatologists have experienced a profound change in their attitude towards anthropoid apes after making eye contact with one for the first time. The spark across the species barrier is never forgotten. Behind the ape's eyes, one can feel a powerful personality that resembles our own, both emotionally and mentally.

eled. Had bonobos been known earlier, reconstructions of human evolution might have emphasized sexual relations, equality between males and females, and the origin of the family, instead of war, hunting, tool technology, and other masculine fortes. Bonobo society seems ruled by the "Make Love, Not War" slogan of the 1960s rather than the myth of a bloodthirsty killer ape that has dominated textbooks for at least three decades.

ARE WE KILLER APES?

In 1925, Raymond Dart announced the discovery of *Australopithecus africanus,* a crucial missing link in the human fossil record. This bipedal hominid with apelike features brought the human lineage considerably closer to that of the apes than previously held possible. It also provided the first indication that Charles Darwin had been correct in suggesting Africa, rather than Asia or Europe, as the cradle of humanity.

On the basis of evidence encountered at the discovery site, Dart speculated that *Australopithecus* must have been a carnivore who ate his prey alive, dismembering them limb from limb, slaking his thirst with their warm blood. The killer-ape myth is the science writer Robert Ardrey's dramatization of these and other ideas, including the proposition that war derives from hunting, and that cultural progress is impossible without aggressivity. The renowned ethologist Konrad Lorenz added that whereas "professional" predators, such as lions and wolves, evolved powerful inhibitions keeping them from turning their weaponry against their own kind, humans have unfortunately not had time to evolve in this direction. Descended from vegetarian ancestors, we became meat-eaters almost overnight. As a result, our species lacks the appropriate checks and balances on intraspecific killing.

It has been suggested that the tremendous appeal of this scenario had more to do with the genocide of World War II than with fossil finds. Confidence in human nature was at a low after the war, and the popularizations of Ardrey and Lorenz merely reinforced the misanthropic mood. In *A View to a Death in the Morning,* Matt Cartmill summarizes the impact of the by now antiquated idea that the lust to kill has made us what we are:

> During the 1960s, the central propositions of the hunting hypothesis—that hunting and its selection pressures had made men and women out of apelike ancestors, instilled a taste for violence in them, estranged them from the animal kingdom, and excluded them from the order of nature—became familiar themes of the national culture, and the picture of *Homo sapiens* as a mentally unbalanced predator, threatening an otherwise harmonious natural realm became so pervasive that it ceased to

> provoke comment. . . . Millions of moviegoers in 1968 absorbed Dart's whole theory in one stunning image from Stanley Kubrick's film *2001*, in which an australopithecine who had just used a zebra femur to commit the world's first murder hurls the bone gleefully in the air—and it turns into an orbiting spacecraft.[1]

Ironically, it is now believed that *Australopithecus*, rather than having been a predator himself, was a favorite food for large carnivores. The damage to fossil skulls, which Dart interpreted as evidence for club-wielding man-apes, turns out to be perfectly consistent with predation by leopards and hyenas. In all likelihood, therefore, the beginnings of our lineage were marked more by fear than ferocity.

BONOBOS AS MODELS

Bonobos are not on their way to becoming human any more than we are on our way to becoming like them. Both of us are well-established, highly evolved species. We can learn something about ourselves from watching bonobos, though, because our two species share an ancestor, who is believed to have lived a "mere" six million years or so ago. Possibly, bonobos have retained traits of this ancestor that we find hard to recognize in ourselves, or that we are not used to contemplating in an evolutionary light.

Not too long ago, a much more distant relative, the savanna baboon, was regarded as the best living model of ancestral human behavior. These ground-dwelling primates are adapted to the sort of ecological conditions that proto-hominids must have faced after they descended from the trees. The baboon model was largely abandoned, however, when it became clear that a number of fundamental human characteristics are absent or only minimally developed in them, yet present in chimpanzees. Cooperative hunting, food-sharing, tool use, power politics, and primitive warfare have been observed in chimpanzees, who are also capable of learning symbolic communication, such as sign language, in the laboratory. Moreover, these apes recognize themselves in mirrors—an index of self-awareness for which there is thus far little or no evidence in monkeys. Like us, of course, chimpanzees belong to the Hominoidea, a branch that split off long ago from the rest of the primate tree. They are thus genetically much closer to us than are baboons.

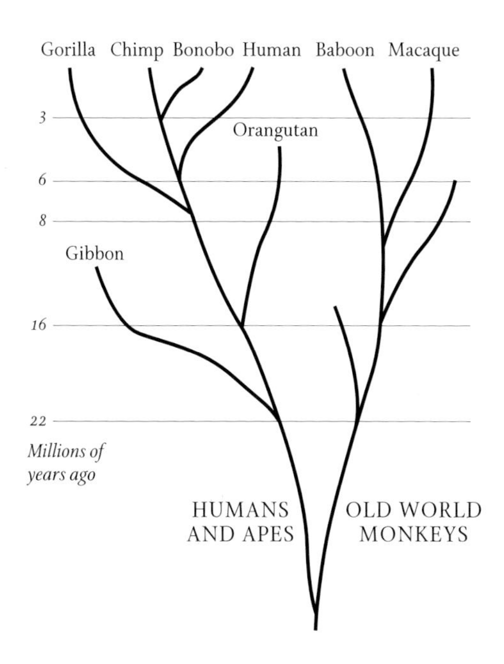

About 30 million years ago, the Old World primate lineage split into two branches: the monkeys and the hominoids. The second branch produced the common ancestor of humans and apes. The human lineage split off an estimated 6 million years ago—well before the split between bonobos and chimpanzees. Thus, neither ape can be considered closer to us than the other. This evolutionary tree is based on comparisons of DNA molecules, the carriers of genetic information.

Whereas selection of the chimpanzee as the touchstone of human evolution represented a great improvement over the baboon, one aspect of the models did not need to be adjusted: male superiority remained the "natural"

Some behavioral scientists had already suspected that bonobos were different before the species was officially discovered, in 1929. This grainy photograph, taken between 1911 and 1916 at the Amsterdam Zoo, shows two apes who at the time were both thought to be chimpanzees, named Mafuca (left) and Kees. A Dutch naturalist, Anton Portielje, wrote that Mafuca might well represent a new species. We now recognize him by his small head, black ears, and long hair as a bonobo. Mafuca was the most popular animal at the zoo. (Photo courtesy of Natura Artis Magistra, *Amsterdam.)*

state of affairs. In both chimpanzees and baboons, males are conspicuously dominant over females. In baboons, males are not only twice the size of females, they are equipped with canine teeth as formidable as a panther's, whereas females lack such weaponry. Sexual dimorphism may be less dramatic in the chimpanzee, but in this species, too, males reign supreme, and often brutally. It is extremely unusual for a fully grown, healthy male chimpanzee to be dominated by a female.

Enter the bonobo, which is best characterized as a female-centered, egalitarian primate species that substitutes sex for aggression. It is impossible to understand the social life of this ape without attention to its sex life: the two are inseparable. Whereas in most other species, sexual behavior is a fairly distinct category, in the bonobo it has become an integral part of social relationships, and not just between males and females. Bonobos engage in sex in virtually every partner combination: male-male, male-female, female-female, male-juvenile, female-juvenile, and so on. The frequency of sexual contact is also higher than among most other primates.

The bonobo's rate of reproduction is low, however. In the wild, it is approximately the same as that of the chimpanzee, with single births to a female at intervals of around five years. This combination of sexual appetite and slow reproduction sounds familiar, of course: nonreproductive sex is a prominent trait of our own species.

If the sole purpose of sex is procreation, as some religious doctrines would have it, why has the average size of families in industrialized nations dropped to fewer than two children, despite the fact that countless human couples in those countries copulate regularly? Perhaps they do so because it feels good, hence tends to become addictive. Yet this automatically raises the question: Why does it have this effect on people? After all, most other animals restrict their mating activity to a particular season or a couple of days in their ovulatory cycles; they do not seem to feel any sexual needs divorced from reproduction.

The bonobo, with its varied, almost imaginative, eroticism, may help us see sexual relations in a broader context. Certain aspects of human sexuality, such as pleasure, love, and bonding, tend to be overlooked by reproduction-oriented ideologies. The possibility that these aspects have characterized our lineage from very early on has serious implications, given how often moralizing relies on claims about the naturalness or unnaturalness of behavior: what is natural is generally equated with what is good and acceptable. The truth is that if bonobo behavior provides any hints, very few human sexual practices can be dismissed as "unnatural."

Because the role of sex in society is such a loaded and controversial issue, scientists have tended to downplay this side of bonobo behavior, whereas the few journalists who have written about the species have naturally hyped it. In this book, I hope to strike a balance: I intend to give the topic the attention it deserves, without reducing bonobos to the lustful satyrs that our closest relations once were considered to be. Sexual encounters of the bonobo kind are strikingly casual, almost more affectionate than erotic. If the apes themselves are so relaxed about it, it seems inappropriate for us to give in to typically human obsessions. In addition, there is a lot more to bonobo natural history than sex. The entire social organization of the species is fascinating, as is its mode of communication, raising of offspring, remarkable intelligence, and status in the wild. The whole creature deserves attention, not just part of it.

In the past few years, many different strands of knowledge have come together concerning this most enigmatic ape. The findings command attention, as the bonobo is just as close to us as its sibling species, the chimpanzee. According to DNA analyses, we share over 98 percent of our genetic material with each of these two apes. And not only are they our nearest relatives; we are theirs! That is, the genetic makeup of a chimpanzee or bonobo matches ours more closely than that of any other animal, including other primates, such as gorillas, traditionally thought of as closer to them than to us.

No wonder Carl Linnaeus, who imposed the taxonomic division between humans and apes, regretted his decision later in life. The distinction is now regarded as wholly artificial. In terms of family resemblance, only two options exist: either we are one of them or they are one of us.

WHAT'S IN A NAME?

Years ago, when the conservator of mammals at the Amsterdam Zoological Museum happened to dust off the stuffed remains of an ape named "Mafuca," he immediately recognized its bonobo features despite the label, which said it was a chimpanzee. During Mafuca's short life, from 1911 through 1916, bonobos were not yet recognized as a separate species, even though a few keen observers already had an inkling of the difference.

In 1916, a perceptive Dutch naturalist, Anton Portielje, speculated in a guide to the Amsterdam Zoo that the hugely popular Mafuca might represent a new primate species. A few years later, Robert Yerkes, the American pioneer of ape research, contrasted "Prince Chim," an individual now known to have been a bonobo, with a chimpanzee, noting: "Complete descriptions of the physique of the two animals might suggest the query as to whether they were both chimpanzees."[2] For all intents and purposes, therefore, the species distinction between bonobo and chimpanzee ought to be credited to behavioral scientists such as Portielje and Yerkes.

It was only when anatomists reached the same conclusion, however, that the world paid attention. The distinction, first made in 1929, carried tremendous weight: the bonobo became one of the last large mammals to be known to science. Rather than in a lush African setting, the historic discovery took place in a colonial Belgian museum following the inspection of a skull that, because it was undersized, was thought to have belonged to a juvenile chimpanzee. In immature animals, however, the sutures between skull bones ought to be separated, whereas in this specimen they were fused. Concluding that it must have belonged to an adult with an unusually small head, Ernst Schwarz, a German anatomist, declared that he had stumbled upon a new subspecies of chimpanzee. Soon the differences were considered important enough to elevate the bonobo to the status of an entirely new species, officially classified as *Pan paniscus.*

Even though Schwarz's name became officially associated with the species—the sort of honor biologists are willing to die for—a far more detailed description was provided, in 1933, by Harold Coolidge, an American anatomist. Half a century later, Coolidge challenged Schwarz's priority. At an international conference of primatologists in 1982, he claimed that he himself had been the first to notice the unusual skull at the museum. In his excitement he had shown it to the museum director, who allegedly told his friend Schwarz two weeks later. Schwarz wasted no time making the discovery public in an obscure journal published by the museum. "I had been taxonomically scooped!" exclaimed Coolidge at the symposium. Unfortunately, Schwarz's side to this story remains unknown: the accusation came after his death.

Oddly enough, the bonobo's genus name, *Pan,* derives from the Greek god of flocks, shepherds, and woods, who had a human torso, but the legs, beard, ears, and horns of a goat. Playfully lecherous, Pan loved to chase the nymphs and played the shepherd's flute, an obvious phallic symbol. The suffix to the species name of the bonobo, *paniscus,* qualifies it as diminutive. The other member of the same genus, the chimpanzee, carries the species name *troglodytes,* or cave dweller. So we are dealing with rather peculiar epithets for animals adapted to the trees, with the bonobo being labeled a small herder deity and the chimpanzee a grotto herder deity.

Since the bonobo and chimpanzee are close relatives, and since the latter is more familiar, the two species are sometimes taken together as two kinds of chimpanzee. Thus, the bonobo is also known as the "bonobo chimpanzee" or "pygmy chimpanzee." Unfortunately, this usage has forced the name "common chimpanzee" upon the chimpanzee—a questionable label for an endangered animal. Furthermore, some scientists object to "pygmy chimpanzee" as inaccurate (there is considerable overlap in size between chimpanzees and bonobos), as well as making it sound too much as if the bonobo is merely a smaller version of its congener. Others, in turn, say the name "bonobo" is meaningless and probably derives from a misspelling on a shipping crate of "Bolobo," a town in Zaire.

The label "bonobo" has stuck, though, not least because it respects its bearer as a fully distinct species, rather than as, so to speak, the poor man's miniature chimp. In addition, "bonobo" has a happy ring to it that befits the animal's nature. Primatologists acquainted with its behavior have even jokingly begun to employ the name as a verb, as in "We're gonna bonobo tonight." (The meaning of this expression will be left to the reader's imagination!)

Robert Yerkes, a pioneer of ape research, with two juvenile apes: a female, named Panzee (left), and a male, named Prince Chim (right). Yerkes, too, had an inkling of the special status of the bonobo before the species was distinguished. Today there is no doubt that Chim, who died of pneumonia in 1924, was a bonobo. Yerkes concluded his book Almost Human *with a moving tribute to little Chim, whom he saw as an intellectual genius quite distinct in behavior and temperament from all other apes. (Photo by Lee Russell, 1923, courtesy of the Yerkes Regional Primate Research Center.)*

To complete these notes on the discovery of the last ape, it has recently come to light that, ironically, the bonobo may have been known to science longer than any other great ape. The earliest accurate description of an ape was produced, in 1641, by Nicolaas Tulp, a Dutch anatomist of great repute, immortalized in Rembrandt's *The Anatomy Lesson.* The ape cadaver that Tulp dissected resembled a human body so closely in its structural details, musculature, organs, and so on, that he commented that it would be hard to find one egg more like another. Although Tulp baptized his specimen an Indian satyr, adding that the local people called it an "orang-outang," it had come straight from Africa. Only its name came from the East Indies (in Malay *orang hutan* means "man of the forest").

Tulp's gravure, faithfully replicated over and over in books of the seventeenth and eighteenth centuries, appears to show a female chimpanzee. At least this was the consensus until a British primatologist, Vernon Reynolds, asserted that Tulp's satyr could very well have been a bonobo. Reynolds's chief argument was that the original drawing shows a cutaneous connection between the second and third digits of the ape's right foot. Such "webbing" between toes is much more common in bonobos than in chimpanzees. Furthermore, Tulp's specimen was known to have originated in Angola. Although no bonobos live there today, Angola is south of the Zaire River. This immense, at times more than one-kilometer-wide water barrier currently fully separates chimpanzees, to the north, east, and west, from bonobos, to the south.[3]

FIRST IMPRESSIONS

Yerkes greatly admired his bonobo's character and intelligence, writing: "I have never met an animal the equal of Prince Chim in approach to physical perfection, alertness, adaptability, and agreeableness of disposition."[4]

Much has been made of this opinion of one of the greatest authorities on ape psychology. Before accepting Yerkes's enthusiasm for Chim as a blanket statement about the species, however, we should realize that the scientist seriously underestimated his subject's age. The slight build of the bonobo led him to believe that Chim was only three years old, whereas a postmortem inspection by Coolidge indicated an age closer to six. In the same way that a child twice the age of another is mentally far ahead, Chim may have come across as brilliant compared to the chimpanzee, Panzee, with whom he was raised. Moreover, Panzee suffered from tuberculosis, another serious disadvantage compared to the healthy Chim. Yerkes himself fully realized the limitations of his comparison, stating that intelligence, temperament, and character very much depend on physical constitution.

Unfortunately, these reservations are rarely mentioned when Yerkes's high regard for Chim is cited in support of claims that bonobos are extraordinarily intelligent. There is no doubt in my mind that they are, but whether their intelligence exceeds that of other apes remains an open question. Simian IQs are about as contentious an issue as human IQs. For one thing, there is great individual variability: comparing a few bonobos with a few chimpanzees is not going to tell us much. I know some exceptionally bright anthropoids, but certainly not all of them are bonobos. At this point it is not at all clear in which cognitive areas, if any, the bonobo systematically outshines other apes.

The first study of substance comparing bonobos and chimpanzees was carried out in the 1930s at the Hellabrunn Zoo in Munich. It took Eduard Tratz and Heinz Heck until after World War II to publish their findings, based on an inspection of the preserved bodies of three apes and film footage collected during their lives: terrified by the city's bombardment during the war, all three bonobos had died of heart failure.[5] Tratz and Heck's eight-point list of behavioral differences between the two *Pan* species still stands as the first outline of the areas of greatest contrast: sexual behavior, intensity of aggression, and vocal expression. Here follows their list in slightly compressed form:

1. Bonobos are sensitive, lively, and nervous, whereas chimpanzees are coarse and hot-tempered.
2. Bonobos rarely raise their hair; chimpanzees often do so.
3. Physical violence almost never occurs in bonobos, yet is common in chimpanzees.
4. Bonobos defend themselves through aimed kicking with their feet, whereas chimpanzees try to pull attackers close to bite them.
5. The bonobo voice contains *a* and *e* vowels, whereas the chimpanzee uses more *u* and *o* vowels.
6. Bonobos are more vocal than chimpanzees.
7. Bonobos stretch their arms and shake their hands when calling, whereas chimpanzees do not.
8. Bonobos copulate *more hominum* and chimpanzees *more canum.*

Chimpanzees are loud! An adult female throws a tantrum after she has been rejected by another female. She screams piercingly while spasmodically hitting herself. Expressions of frustration and hostility, as well as of joy and excitement, are conspicuous in this species known for its boisterous, extroverted, and querulous temperament. (Photo by Frans de Waal.)

Given what we know now, points 1 through 4 are undoubtedly correct. Even though the difference in aggressivity is one of degree only, it cannot be denied that the treatment to which chimpanzees occasionally subject one another, including biting and full-force hitting, is rare among bonobos. Chimpanzees also erect their hair at the slightest provocation, pick up a branch, and challenge and intimidate anyone perceived as weaker than themselves: they are very much into status. By bonobo standards, the chimpanzee is a wild and untamed beast, or as Tratz and Heck put it: "The bonobo is an extraordinarily sensitive, gentle creature, far removed from the demoniacal primitive force [*Urkraft*] of the adult chimpanzee."[6]

As regards point 5, Blanche Learned's pioneer (albeit unwitting) comparison of vocal repertoires is worth noting. Before the species difference was established, she listened with a musical ear to Yerkes's two apes, Chim and Panzee. According to my calculations from Learned's phonetic transcriptions of hundreds of vocalizations, Chim mostly uttered *a* (48%), *ae* (38%), and *oo* (10%) sounds, whereas Panzee mostly uttered *oo* (68%), *o* (12%), and *oa* (7%) sounds. There is indeed no quicker way to distinguish the two ape species than by their voices. When Heck, who was the director of Hellabrunn Zoo, first heard bonobo calls coming out of a cloth-covered crate, he was convinced that he had received the wrong animals. Their calls are so high-pitched and penetrating that they do not even remind one of the typical drawn-out "huu . . . huu" hooting of the chimpanzee. The difference in timbre between the voices of the two species may well be of the same magnitude as that between a small child and a grown man.

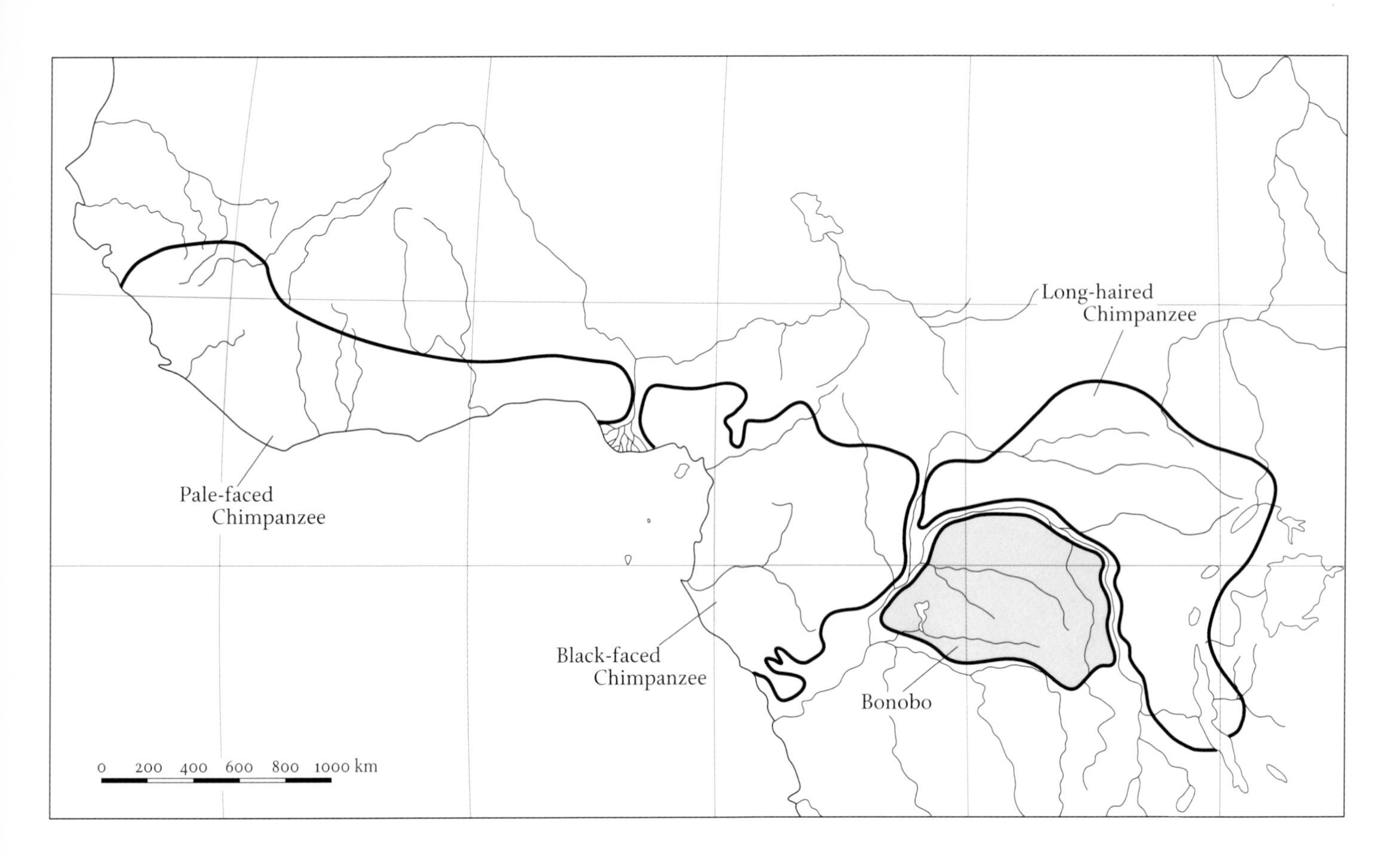

Map of equatorial Africa showing the probable distribution of members of the genus Pan, *around 1900. Because of habitat destruction, the current distribution is considerably more fragmented. There exist three subspecies of chimpanzee: the masked or pale-faced chimpanzee in West Africa, the black-faced chimpanzee in Central Africa, and the smaller, long-haired chimpanzee in East Africa. Bonobos are entirely restricted to an area south of the Zaire River. This map is based on Dirk Thijs van den Audenaerde's summary of historic locality records at the Tervuren Museum, Belgium.*

It is also true that bonobos tend to gesticulate when calling, and that vocal activity among them is high. Bonobos are excitable creatures who frequently "comment" on minor events around them through high-pitched peeps and barks. Even if most of these vocalizations are noticeable only at close range, one definitely hears more vocal exchange in a group of bonobos than in a group of chimpanzees. Chimpanzees call when seriously alarmed, aroused by food, or in order to intimidate one another. Few animals can produce the din characteristic of chimpanzees, but much of it occurs on well-circumscribed occasions.

The final point concerns sexuality. Because Tratz and Heck wrote before the sexual revolution, they felt the need to wrap their shocking findings in Latin. In those days, face-to-face copulation was regarded as uniquely human; a cultural innovation reflecting the dignity and sensibility separating the human race from "lower" life forms. The two zoologists claimed, however, that whereas chimpanzees mate like dogs (*more canum*), bonobos follow the human pattern (*more hominum*). They added the important observation that the genitals of female bonobos seem adapted to this position: the vulva is situated between their legs rather than oriented to the back, as is the case in chimpanzees.

To this day, both academic and popular writers perpetuate ridiculous claims about human mating patterns, penis size, and general sexiness.[7] The primary reason for overlooking the considerable early knowledge about bonobos must have been that most of it was unavailable in English. Who browses through journals such as *Säugetierkundliche Mitteilungen*? Apart from their role in the naming game (they were the first to propose "bonobo"), Tratz and Heck were ignored and forgotten by the scientific community. Another overlooked work is an admirably detailed investigation at three European zoos by Claudia Jordan, whose 1977 dissertation, "Das Verhalten zoolebender Zwergschimpansen" (The Behavior of Zoo-living Pygmy Chimpanzees), contains virtually all of the basic behavioral information presented as new discoveries in the literature of subsequent years.

A second reason that little attention was paid to some of the early studies was the tendency to dismiss unusual behavior in zoo animals as artifacts of captivity. Could it be that bonobos act so grotesquely because they are bored to death, or under human influence? We know now that, except under extreme conditions, the effects of captivity on behavior are less dramatic than used to be assumed. Whatever the conditions under which *other* primates are kept, they never act like bonobos. In other words, it must be something in the species, rather than in the environment, that produces the bonobo's characteristic behavior. It was only when fieldwork got off the ground, however, that the behavior-as-artifact explanation could be put to rest. Research in the bonobo's natural habitat validated rather than contradicted the pioneering observations of Yerkes, Tratz and Heck, Learned, Jordan, and others.

In 1974, Alison and Noel Badrian, a young couple of Irish and South African extraction, bravely entered the remote jungles of northern Zaire on their own, without financial backing. They established a study site in Lomako Forest, which is still in use today, although observation has been discontinuous and conducted by a number of different scientists. The other main study site in Zaire, established in the same year, has known much greater continuity and has, as a result, become the dominant source of information about wild bonobos. This site, named Wamba, was founded by Takayoshi Kano of Kyoto University, in Japan, after a five-month survey of the distribution of Zaire's bonobo population. Transportation by other means being virtually impossible in this region, Kano traveled enormous distances on foot and by bicycle.

These and other dedicated fieldworkers have advanced our knowledge of bonobo behavior by giant strides, confirming the significance and richness of these apes' sexual behavior and putting their social organization in the context of the ecological background to which it is adapted: the swampy rain forest covering the flat basin of the Zaire River. Because wild bonobos are extremely shy, it takes a long time to habituate them to human presence. At Wamba, this problem was solved by a technique widely employed with Japanese macaques in the investigators' home country: food provisioning. By planting a few hectares of sugarcane near their range, Kano was able to entice bonobos out of the forest. At Lomako, such techniques have never been employed. The Lomako site has therefore something unique to offer: a look at the ranging and foraging patterns of bonobos undisturbed by human provisioning.

Despite the establishment of Wamba, Lomako, and a handful of other field sites, bonobo research still lags far behind that on chimpanzees in both scope and intensity. Over recent years, however, interest has grown rapidly, not least because bonobos seem to present a mirror-image of the traditional picture of our primate relatives as male-dominated and violent. As a feminist journalist for a nature magazine once put it to me: "Bonobos are our only hope!" An ideological interest in the species may not sound desirable to most scientists, yet so long as it leads to scholarly, honest, and rigorous study, I do not see much wrong with it. As a result of continued research, current impressions and theories will either be confirmed or require revision, and we shall gain a deeper understanding of why bonobos evolved the sort of society that they live in.

In the meantime, captive bonobos have become more attractive for behavioral studies: zoo colonies now include more individuals in more naturalistic enclosures than the single individuals or small groups of the past. In addition, the apes live longer than before. Bonobos are extremely susceptible to respiratory disease: they used to survive only a couple of years in captivity. With greater care and better nutrition, there now are bonobos aged twenty, thirty, or older in zoos and research institutions. The development towards improved survival and larger social groupings began at the San Diego Zoo, where I conducted my own research. This zoo started out very modestly, in the early 1960s, with a single pair of bonobos: Kakowet and Linda. These two were so prolific that they produced the greatest number of children and grandchildren known of any bonobo couple, captive or wild, in the world. Part of the reason was that every newborn was taken away to the zoo nursery; this allowed Linda to skip the long nursing period and deliver at unusually short intervals: ten children in fourteen years.

Not that this is a desirable procedure! Many of Linda's newborns were featured on Johnny Carson's late-night television show, and I feel they would have been better off with a little less fame and a little more motherly love. Nowadays, zoos, including the San Diego Zoo, do everything in their power to keep mother and infant together.

Linda is still alive (estimated to be around forty, she now lives with one of her adult daughters at the Milwaukee County Zoo), but Kakowet died years ago. Stories about the patriarch of zoo bonobos abound. According to one, Ernst Schwarz was overjoyed to hold Kakowet when he was still a small infant (his name derives from the French word for peanut, *cacahuète,* because he was so incredibly tiny).[8] Having conducted all of his taxonomic work with museum skeletons and skins, Schwarz had never met a live bonobo. Standing with the ape on his arm, the German anatomist was greeted by a woman who said: "So, you're the man who named that funny little monkey." A shocking thing to say to someone so familiar with the distinction between monkeys and apes![9]

Now that captive bonobos have become more interesting for students of social behavior, and fieldwork is growing in both quality and quantity, we are in a better situation than ever before to summarize this ape's social life. Our knowledge is far from perfect, but we know enough to drag the bonobo out of the obscure corner in which primate specialists have been debating its peculiarities among themselves. Its behavior is bound to overthrow a number of cherished assumptions about the course of human evolution. In addition, the species is fascinating in itself; it fully deserves a place in the public mind alongside our better-known ape relations.

ALL IN THE FAMILY

Bonobos—here a juvenile male at the base of a giant tree—are the least studied, hence least understood, of the four great ape species. The extreme remoteness of their natural habitat in Central Africa is partly to blame, as is the political instability in this region. Moreover, the long-standing assumption that bonobos resemble chimpanzees in behavior failed to inspire fieldwork. We now know, however, that bonobos have a dramatically different social organization. They fully deserve their species status as Pan paniscus *and are drawing increasing attention from science.*

The orangutan (Pongo pygmaeus) *is the most flamboyantly colored member of the hominoid family. The only Asian great ape, it roams the steamy jungles of Borneo and Sumatra. The species is marked by dramatic sexual dimorphism: males are about twice the size of females. In addition, male orangutans develop cheek pads of fibrous tissue that enlarge their faces (compare the adult male, above left, with the adult females, above right and opposite). Full-grown males are extremely intolerant of one another: they utter characteristic roars to warn off intruders and attract mates. Orangs live solitary lives: except for the occasional consortship between a male and a female, adults travel on their own. This ape's anatomy (long arms, hook-shaped hands and feet, short legs) suggests a long history of arboreal adaptation.*

The gorilla (Gorilla gorilla) *is the King Kong of Hollywood fame. This ape is the undisputed heavyweight of the primate order, with males weighing up to 180 kg (400 pounds). As in orangutans, males and females differ greatly in size. Gorillas have an undeserved reputation for being ferocious. Only in extreme cases, such as when they defend their family group against hunters, do they attack people. They really are gentle giants. Largely terrestrial, they travel in permanent groups of usually fewer than ten individuals: a fully adult male—the so-called silverback—several adult females, and their immature offspring. Once a male has established a harem, he tends to stay with it for life. Serious aggression breaks out only when a silverback leader is challenged by another male, who may take over the unit, sometimes killing infants in the process. Both photos show lowland gorillas: a captive adult female (above) and a silverback munching on swamp vegetation in Kahuzi Biega National Park, Zaire (opposite).*

Chimpanzees (Pan troglodytes) *are the most familiar apes. Common in zoos, they figure prominently in studies of intelligence, and their behavior in the wild is widely known. The difference in size between the sexes is much smaller in chimpanzees than in gorillas and orangutans; in this respect they resemble bonobos (and humans). Furthermore, both bonobos and chimpanzees live in so-called fission-fusion societies (see p. 63). But otherwise there are major differences: chimpanzee males are more dominance-oriented and "political" than bonobo males, and chimpanzee females have less social influence than their bonobo counterparts. The adolescent male chimpanzee above still has several years to go before he will dominate the females of his species, but in general adult males dominate females of all ages. The female shown opposite carries an infant with a light-colored face, a feature that distinguishes chimpanzees from other African apes; bonobo and gorilla infants are born dark-faced.*

CHAPTER 2

TWO KINDS OF CHIMPANZEE

The new is always contrasted with the common: the main thrust of discussions about bonobos is how they compare with chimpanzees. Chimpanzees have been studied in captivity since the pioneering work, in the 1920s, of the German psychologist Wolfgang Köhler, who tested their tool-use capabilities, and Robert Yerkes, who studied their temperament and intelligence. Moreover, in addition to the well-known observations in Gombe National Park by the British primatologist Jane Goodall, several other ongoing field projects on chimpanzees provide us with a solid knowledge base.

Comparing bonobo behavior with what we know about chimpanzees is complicated, however, by the recent discovery of considerable "cultural" variation. The idea that chimpanzees act essentially the same everywhere is rapidly being abandoned, and the same variability no doubt exists among other apes. The differences to be discussed here are in addition rather superficial relative to the similarities. Both ape species are large mammals adapted to an arboreal life, both spend most of their time foraging for fruit, both have long-lasting mother-offspring bonds, and both know male-male competition over status.

An adult male bonobo in the equatorial dusk.

The two apes are fundamentally similar in anatomy, behavior, and natural history—they were rightly placed together in the same genus.

Nevertheless, I shall take this common ground for granted and emphasize the differences. We often do the same when comparing ourselves with other animals: we ignore the shared characteristics and zoom in on the differences. This is a common strategy to identify our special place in the natural world; we can follow the same strategy to define the bonobo's place.

Australopithecus afarensis *lived about 3.5 million years ago. This diminutive hominid had a chimp-sized brain but already walked upright, which indicates that in the course of human evolution bipedal locomotion came before brain expansion. We know too little about the behavior of this ancestor to be sure that males and females walked together as couples, as suggested by this reconstruction, but footprints found in Laetoli, Tanzania, confirm a well-developed bipedal stride. (Lucy diorama, American Museum of Natural History, New York.)*

LIVING LINKS

The bonobo cannot be distinguished from the chimpanzee on the basis of size. A recent survey of virtually all captive bonobos in the world by an American anthropologist, Amy Parish, put the weight of the average male at 43 kg (95 pounds) and of the average female at 37 kg (82 pounds). This is equal to or heavier than the smallest subspecies of chimpanzee, but below the weight of the other two chimpanzee subspecies.

Parish also found that sexual dimorphism (the size difference between adults of both sexes) may be slightly reduced in bonobos. In chimpanzees, the weight of females is on average around 80–84 percent that of males. Only in the Gombe population, with its unusually small chimpanzees, is the sex difference greater, with females 71–75 percent of male weight. In bonobos and humans, on the other hand, females weigh around 85 percent of what males weigh. But even if males and females differ relatively little in size—particularly in comparison to such dimorphic hominoids as the gorilla and orangutan, among which females may be less than half the average male size—a weight disadvantage of 15 percent is still quite significant if it comes to a fight. Bonobo males are not only heavier but also more muscular than females, and they are equipped with long canines, which the females lack. Clearly, from the build of the sexes alone, no one would have predicted an egalitarian society.

The weight overlap between bonobos and chimpanzees notwithstanding, the former species immediately strikes us as more gracile and elegant. Even chimpanzees would have to admit that bonobos have more style. The bonobo's body is slim and slender, and the head is small, on a thin neck and narrow shoulders. The lips are reddish in a black face, the ears are small, and the nostrils are almost as wide as a gorilla's. Bonobos also have flatter, more open faces, with higher foreheads than chimpanzees, and—to top it all off—they all have the same hairstyle, with long, fine, black hair neatly parted in the middle.

The main difference between the species is in their body proportions. The chimpanzee has a large head, thick neck, and broad shoulders, whereas the bonobo's upper body is that of a featherweight. Instead, bonobos have strikingly elongated legs. The result is that, when knuckle-walking on all fours, a chimpanzee's back slopes down from the powerful shoulders, whereas a bonobo's back remains fairly horizontal because of the elevated hips. When standing or walking upright, the bonobo's back seems to straighten better than the chimpanzee's, giving the former a very humanlike posture. Adrienne Zihlman, an American physical anthropologist, has argued that of all the living apes, the weight distribution of the bonobo is closest to that of the prehistoric African "ape-men," or Australopithecines. In some apes, such as the quadrupedal orangutan, lower and upper limbs weigh about the same. Habitual bipeds, such as ourselves, on the other hand, have shifted much weight towards the lower limbs. Between these extremes, the bonobo is more like us than like some of the other apes. Zihlman takes this to mean that the human lineage may have evolved from a common ancestor who looked a lot like the long-legged bonobo.

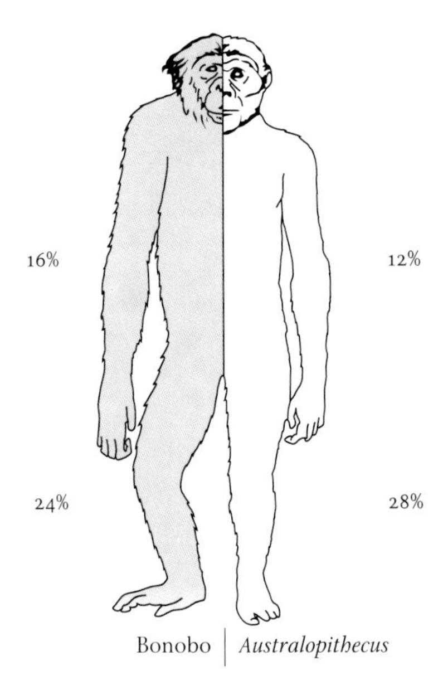

When Adrienne Zihlman measured the limbs of the australopithecus *fossil, "Lucy," she calculated that the legs must have been more than twice as heavy as the arms. This weight distribution approaches that of specialized bipeds, such as ourselves: our legs weigh nearly four times as much as our arms. Comparing extant apes with Lucy, Zihlman found the bonobo's body proportions to be the most similar. The above drawing illustrates her point. Lucy's arms take up 12 percent of her body weight and her legs 28 percent, whereas for bonobos these figures are, respectively, 16 percent and 24 percent. Could it be that the common ancestor of humans and apes was built like a bonobo, hence preadapted for bipedal locomotion? (Based on an original drawing by Carla Simmons, courtesy Adrienne Zihlman.)*

This suggestion should not be confused with a claim that we are more closely related to the bonobos, which is simply impossible, given that the split between the two *Pan* species occurred millions of years after we split off from them: bonobos and chimpanzees are equidistant to us. No, what Zihlman proposes is that the bonobo might be the best model of the so-called missing link. This view echoes Coolidge's oft-cited conclusion that the bonobo "may approach more closely to the common ancestor of chimpanzees and man than does any living chimpanzee."[1]

In other words, since the days when our mysterious common ancestor walked the earth—an estimated six million years ago—the bonobo's body plan may have undergone less modification than the chimpanzee's. Our own species evolved plenty of new characteristics, such as bipedal locomotion, huge brains, and loss of body hair. Evolutionary change in the chimpanzee may have been prompted by a need to adapt to half-open, dryer habitats, such as savannas and woodlands. Bonobos, on the other hand, probably never left the protection of the rain forest; at present, they are entirely restricted to wet equatorial regions. For this reason, Takayoshi Kano believes that the bonobo may have experienced little need for transformation, and hence have retained more ancestral characteristics than either humans or chimpanzees. If so, the bonobo would most resemble the prototype of the three modern species.

This is one school of thought; another school, in contrast, maintains that the bonobo underwent quite a few changes. This position is supported by

HOMINIZATION IN THE FOREST

It has long been assumed that the defining feature of our race—our bipedal gait—evolved on the African plains. Perhaps our ancestors adopted an upright posture to look over the tall grass, to free their hands to carry weapons, to reduce the heat absorbed from the burning sun by exposing less of their bodies, or to increase locomotion efficiency. Some scientists believe that walking on two legs takes less energy than walking on all fours when moving across the savanna. A growing body of fossil evidence from shortly after the divergence between our direct ancestors and the apes suggests, however, that the earliest protohominids still dwelled, at least partially, in forest habitats. The transition to bipedal locomotion was probably more gradual than scientists used to think. Our ancestors may have gone through a long intermediate stage in which they walked around on two legs yet also visited the trees.[2]

Two South African anthropologists, Ronald Clarke and Phillip Tobias, recently unearthed four small bones at Sterkfontein that formed part of the left foot of a human ancestor estimated to have lived at least 3 million years ago. The foot had a weight-bearing heel adapted for bipedal locomotion but also an apelike toe that could still grasp and wriggle. A close cousin of the famous "Lucy," this Australopithecus fossil was baptized "Little Foot." It sparked hot debate between scientists who refuse to imagine our forefathers swinging through the trees and others arguing that it is unlikely that a foot suited for climbing would not have been used for that. If Little Foot, as Clarke and Tobias diplomatically put it, did not sacrifice arboreal competence, this was probably because he or she could not survive without it. Perhaps Little Foot still needed to get into the trees to collect fruits or escape predators. Like its ape cousins, it may also have found it safer to sleep in elevated nests rather than on the ground.[3]

The controversy surrounding this fossil has made scientists take another look at the bonobo, which has long legs and is an excellent bipedal walker, but lacks the specific ankle that makes Little Foot, and not the bonobo, one of our direct forebears. When Randall Susman, an American anatomist, documented bonobo locomotion at Lomako, he saw parallels with the locomotion of early hominids. In apes, too, the transition from arboreal to terrestrial life may have occurred little by little. The common ancestor of the African apes achieved this shift through knuckle-walking, a form of locomotion on the ground compatible with

OPPOSITE: Because nonhuman primates climb, brachiate, and walk on all fours, their hands serve a crucial locomotor function. Our hands have been freed from this task, which has allowed our thumbs to become more fully opposable to the fingers than in any other primate. As a result, human hands are ideally suited to transporting objects and using tools. Our feet, however, are exclusively adapted for locomotion. Note the grasping feet of apes and gibbons, with big toes set apart from the other digits, compared with our weight-bearing feet. (Drawings by Adolph H. Schultz.)

the preservation of long fingers needed for climbing. Susman goes on to speculate:

> The earliest hominid ancestor also began to adapt to life on the ground, and like African apes today, it, too, retained for some time the capacity to climb trees. Why early hominids elected a bipedal rather than knuckle-walking (or some other) solution to the problem of terrestriality is perhaps the key question. There is wide divergence of opinion, but most present hypotheses incorporate the need for freeing the hands for some form of carrying. In this regard, it is interesting to note that many who have observed bipedal locomotion in free-ranging chimpanzees (common and pygmy) have noted that bipedality is strongly linked to food carrying or carrying objects during display.[4]

It is entirely possible that not only bipedality but also other aspects of hominization, such as tool-making and cooperative hunting, evolved in the forest before our ancestors began putting these skills to use on the plains. Hedwige and Christophe Boesch defend this thesis based on observations of forest-living chimpanzees.

At first sight, bonobos seem to present contradictory evidence: they live in dense forest, yet tool-making and cooperative hunting have thus far not been documented. Study of this ape may illuminate a different issue, however. Existing models of hominization rely heavily on male bonding and cooperation, whereas there can be no doubt that female bonding is equally prominent in our lineage. This may well be another heritage from our history as forest-dwellers, one that we share with our gracile cousins.

---

unique characteristics of bonobo chromosomes,[5] blood groups, dentition, sexual anatomy, and reproductive physiology. It has been speculated that bonobos evolved through retention of juvenile characteristics into adulthood, a process known as *neoteny*. For example, the smaller skull of the adult bonobo reminded both Schwarz and Coolidge of a juvenile chimpanzee. Bonobos also keep their white tail-tufts, which chimpanzees lose after weaning age. The voices of adult bonobos are as shrill as those of juvenile chimpanzees, and even the frontally oriented vulva is considered a neotenous characteristic, also present in our own species. Neoteny has been called the hallmark of human evolution; it is reflected in our hairlessness, large brains, and general playfulness. So, bonobos may have differentiated themselves from chimpanzees through a process similar to that underlying our own spectacular evolution.

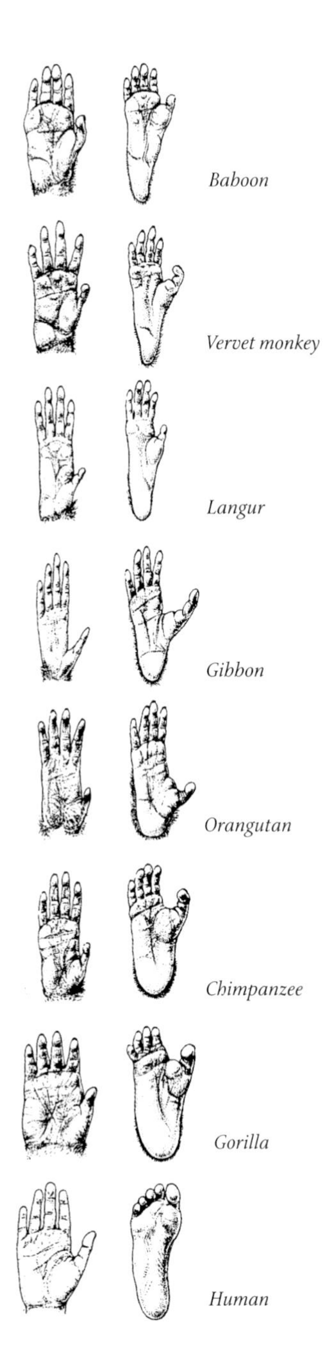

Either way—whether bonobos are the more primitive or the more derived form—they have an interesting story to tell about ourselves.[6]

OPPOSITE: Because of climatic changes, the forestation of the African continent has varied throughout the ages. According to Elizabeth Vrba, a South African paleontologist, lower temperatures and reduced rainfall dramatically shrank the forest cover during glacial periods. Many species, including perhaps our ancestors, may have been forced to adjust to relatively open habitats. If bonobos lived in one of the "rain forest refugia" indicated on the second map from the top, they were buffered from such pressures for change. (Based on map in Donald Johanson, Lucy's Child.*)*

An increasingly important source of information about the past is DNA sequencing. Comparison of human and ape DNA has resulted in a thorough revision of the evolutionary tree, placing the African apes much closer to us than previously held possible. DNA analyses are being refined further and further, and they have recently begun to challenge the view that there are only two species within the *Pan* genus. There exists quite a bit of variation among the three subspecies of chimpanzee, known as masked, black-faced, and long-haired chimpanzees. In terms of DNA characteristics, the western masked chimpanzee seems to stand out: it has been argued that these apes should be considered a separate species. This suggestion is still controversial, though, and for the moment I shall assume that the genus includes only two species.[7]

SMILES AND FUNNY FACES

When studying a new species, the first thing one needs to do is draw up a so-called *ethogram*: a systematic description of its behavior patterns. For the chimpanzee, for example, we have two first-rate ethograms by Jan van Hooff, a Dutch primatologist, and Jane Goodall. Although the first ethogram concerns captive apes and the second wild ones, it is remarkable how much they agree. Evidently, species-typical vocalizations, gestures, facial expressions, and other forms of communication vary little from one setting to another.

Combined with my own knowledge of chimpanzees, I took the above two ethograms as the starting point for a project at the San Diego Zoo. Its aim was to describe the behavior of bonobos in direct comparison with their congeners.[8] At the time, the San Diego colony included ten bonobos, which I observed standing in front of their enclosure speaking my observations into a tape recorder. At moments of great activity, such as at feeding time or when new group members were introduced, I used a video camera. The focus was on signals that convey motivational states, such as hostility, need for contact, sexual urges, and so on. I gathered information on so many behavioral sequences (5,135, to be exact) that it took much longer to enter all this information into a computer than it took to collect it.

Of the over fifty behavior patterns that I distinguished, more than half occurred in both species. These shared patterns were not necessarily exactly the same, but at least similar and equivalent in meaning. For example, both species will stretch out an open hand to a companion to request food, support, or contact. When a mother walks away from her offspring, the juvenile may hold out a hand and whimper, which makes the mother return and retrieve the one left behind. In the bonobo—for whom hands and feet are largely equivalent—the same begging gesture may also be given with an outstretched foot.[9] Further, bonobos often add so-called finger-flexing, in which the four fingers of the open hand are bent and stretched in rapid alternation, making the invitation look more urgent.

One day, two adult males were introduced after a long separation. They both screamed and turned around each other for six minutes without any physical contact. We feared a bloody confrontation (most animals fight when introduced to a relative stranger of the same sex), but Kevin, the younger male, kept stretching out his hand and flexing his fingers, as if beckoning Vernon to come closer. Occasionally, Kevin shook his hands impatiently. Both males had erections, which they presented to each other with legs apart, in the same way that a male invites a female for sex. It was as if each male wanted contact but did not know whether the other could be trusted. When they finally did rush towards each other, instead of fighting, they embraced frontally with broad grins on their faces, Vernon thrusting his genitals against Kevin's. They calmed down right away and happily began collecting the raisins that the caretakers had scattered around. Instead of screaming, they now uttered excited food calls.

The way this brief but tense encounter unfolded is emblematic of the species: the role of genital contact, the intense exchange of signals, and the peaceful ending. Aggressive behavior is not absent in the bonobo, either in captivity or in the field, but it remains mostly mild and restrained compared to the elaborate charging displays for which the chimpanzee is known. A male chimpanzee appears larger than life when he raises his hair, uproots a small tree, and charges about slapping the ground with great force and energy. When he is in this mood, anyone who crosses his path risks a beating. He can keep up the performance for minutes; perhaps the length and vigor of the display informs his fellows about his health and stamina. In the Mahale Mountains National Park, in Tanzania, Toshisada Nishida observed a high-ranking male who had developed a habit of displaying near a riverbed with enormous rocks, which he would dislodge with superhuman strength and roll downhill, producing a thunderous noise, which seemed to impress his rivals.

In comparison, the bonobo male's typical display looks like child's play. He will grab a branch and drag it behind him while making a brief run. There is almost no comparison with the unstoppable steam-engine display enacted by

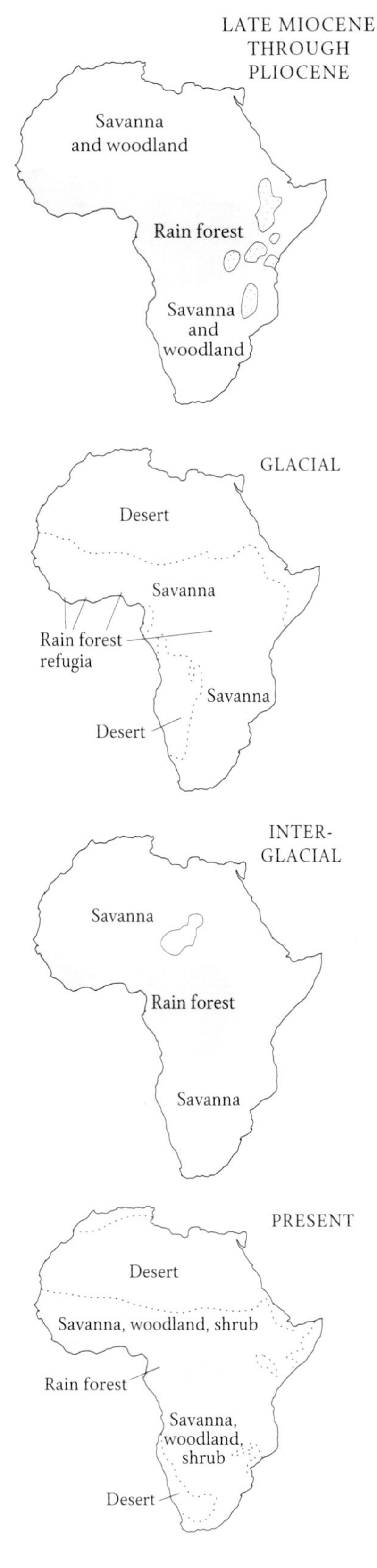

OPPOSITE: This aerial view of current forest and grassland habitats in Central Zaire suggests how historic changes in the forest cover may have created a new niche for apes, including our ancestors. Some populations may have been forced to rely increasingly on savanna resources.

his more robust cousin. Also, bonobos rarely perform the complex confrontations known among chimpanzees, in which one opponent recruits supporters against the other, thus forcing the latter to do the same, until entire sections of society oppose each other on the battlefield. Chimpanzees will go around cajoling their friends to get involved, holding out a hand to one, embracing another. As a result, confrontations may last half an hour or more, involving all sorts of shifting alliances screaming and barking at each other. Bonobos, in contrast, fight chiefly on a one-on-one basis without tactical maneuvers to draw in third parties.

This is not to say that a bonobo placed with chimpanzees would be at a loss about what is going on among them, or that bonobos themselves never form alliances. They are capable of the same kind of interactions, yet rarely engage in confrontations on the grand scale characteristic of the chimpanzee's political system. Chimpanzees also go through elaborate rituals in which one individual communicates its status to the other. Particularly between adult males, one male will literally grovel in the dust, uttering panting grunts, while the other stands bipedally performing a mild intimidation display to make clear who ranks above whom. Overall, communication patterns related to aggression, dominance, and submission are more conspicuous and spectacular in the chimpanzee. This species also seems to invest considerably more energy, both physically and mentally, in politicking: chimpanzees are the Machiavellis of the primate world.

Are we permitted to speak of "politics" in relation to animals other than ourselves? If we follow the social scientist Harold Lasswell, who designed the classical definition of politics as "who gets what, when, and how," there is no reason why the dominance strategies and alliances of nonhuman primates should not be labeled as such. For example, in baboons and chimpanzees, it is not unusual for two males to band together to defeat a third; such alliances determine sexual access to females, hence decide who gets what. As soon as alliances come into play, a higher level of social awareness is required. It becomes of paramount importance to know, not just who your own friends and enemies are, but also who the regular allies of your adversaries are, so that their presence or absence can be taken into account. Success demands keeping track of a large number of social factors, hence the thesis, forwarded by primatologists in the 1950s and 1960s, that social problem-solving is the original function of primate (including human) higher mental faculties.

If chimpanzees indeed engage in the most elaborate power strategies, they might represent the best example of socially driven brain evolution. Does this

mean that the bonobo, being less of a political animal, is less intelligent? Indications are that bonobos resolve the question of "who gets what, when, and how" by different, somewhat subtler means. The chimpanzee tries to gain influence and get the upper hand, whereas the bonobo follows a less power-hungry scheme. After all, there are many ways to get what one wants, and in bonobo society, the accent seems to have shifted towards peaceful settlement of conflicts of interest.

This brings me to the domain in which bonobos excel. If, of the twin concepts of sex and power, the chimpanzee has an appetite for the second, the bonobo clearly has one for the first. The chimpanzee resolves sexual issues with power; the bonobo resolves power issues with sex. It is important to consider bonobo sexual behavior in detail, so much so that I devote all of chapter 4 to it. Suffice it to say here that in sexual matters, chimpanzees are the ones to follow an almost laughably simplistic scheme. All of this was reflected in my ethogram, which naturally reached its greatest richness in the areas in which each species showed most variation. Bonobos have more ways of inviting each other sexually, more ways of engaging in sex, and more different facial expressions and vocalizations associated with sexual intercourse than chimpanzees. The chimpanzee's sex life is rather plain and boring; bonobos act as if they have read the *Kama Sutra*.

By far the most striking contrast in the ethogram, however, concerned vocal communication. As noted by Tratz and Heck, the voices of the two species are so different that it is easy to tell them apart. Not only is the bonobo's voice shriller, the calls are often different. For example, the long-distance call of the chimpanzee is a slowly swelling hooting, whereas the bonobo utters rather nervous, agitated whooping calls. From a distance, it sounds like the yapping of a small dog—or rather many small dogs, since bonobos synchronize these calls to a high degree, producing choruses in which individuals "echo" each other.

An even more remarkable coordination is achieved during aggressive confrontations. Each opponent may vocalize in alternation with the other, exchanging sounds as fast as the back-and-forth of a professional ping-pong match. They seem to be trading information about emotions and intentions. The calls are quite variable: some may be threats, others may betray fear, and yet others express a desire to reconcile. During the introduction between Kevin and Vernon, the two males went through this kind of rapid dialogue in a very intense manner. A spectrographic analysis of their calls indicated that they changed in quality over the course of the encounter but remained similar to each other (a process known as vocal matching), as if the two males gradually converged on a solution to their predicament. Their calls almost never overlapped—that is, the two males responded to each other without interrupting. Based on these observations, and Claudia Jordan's similar description of so-called squeal duels (*Quiekduelle*), I hold it possible that what chimpanzees do by means of visual displays of strength and determination, bonobos do by means of a more "languagelike" exchange of information about internal states. It is not that they are talking the issue over, but there is a certain dialectic and coordinated quality to their vocal contests that is absent in the chimpanzee.

The facial expressions of the two species, on the other hand, are remarkably similar. Pouting lips express a desire for contact or disappointment about some frustrating event. For example, Kalind, an adolescent male in the San Diego colony, was often driven away from the top male's favorite female. Afterwards, Kalind would sit staring into the distance with pouted lips. In this sulking mood, he often caressed his own nipples with rapid movements of his thumbs, perhaps an act of self-reassurance.

During play, the mouth is opened in a relaxed expression, and coarse, guttural panting sounds are heard when partners tickle each other. This "laughing" is the only sound that is virtually indistinguishable in the two species. In an aggressive mood, the lips are compressed in a tense face with frowning eyebrows and piercing eyes. When fearful, on the other hand, the lips are retracted, exposing the teeth in a broad grin. This is confusing to people, because we tend to associate a smile or grin with cheerfulness and affection. Interestingly, this meaning is not entirely absent, because bonobos also grin when they discover a new toy or reach the climax of copulation or masturbation. During my study, Louise, an adult female, would show a wide grin while pirouetting around and around in a nest built with fresh branches and leaves. In all of these cases, the apes gave the impression of being elated and satisfied. Such "pleasure grins" should be distinguished from the more common fearful or nervous baring of the teeth. This may seem a totally contradictory use of the same facial expression, but not if one assumes that fearful situations, such as the sudden approach of a potentially hostile dominant, often call for appeasement. It is this mollifying, friendly quality of the expression that has gradually gained in importance—very prominently so in the smile of our own species (although nervous grins have by no means disappeared), and to a lesser degree in the bonobo.[10]

Finally, I should recount how, when drawing up my ethogram, I found myself in a situation that in retrospect is rather amusing. Each time I worked on a group of juveniles in a spacious, green enclosure, my list of facial expressions grew longer and longer. I had to describe the weirdest expressions, and they were never exactly the same as before. It was impossible to tell what they meant. After a while, it dawned on me that these expressions always occurred in nonsocial situations. They were neither preceded nor followed by particular actions, such as sex or aggression, that might help their interpretation. A young bonobo would be staring at nothing in particular and suddenly go through a pantomime of sucked-in cheeks, bulging upper lips, and rapid jaw movements. Sometimes hands would get involved, for example by pulling a lip sideways or reaching all the way around the back of the head to stick a finger into the mouth from the "wrong" side.

I decided that the bonobos were just amusing themselves with fantasy grimaces that served no communicatory function whatsoever. That they engaged in these "funny faces" is intriguing in itself, since it indicates that they possess voluntary control over their facial musculature. Could not an animal that pulls faces for fun do the same in order to manipulate others? Whatever the implications, these young apes certainly made me see the foolishness of science's obsession with classification. Were they mocking me? Once I realized what their facial acrobatics were all about, I could not suppress the feeling that they occasionally winked at me!

BONOBO BRIGHTNESS

An old Kakowet story at the San Diego Zoo suggests high intelligence. The two-meter-deep moat in front of the bonobo enclosure had been drained for cleaning. After having scrubbed the moat and released the apes, the keepers went to the kitchen to turn on the valve for refilling when all of a sudden Kakowet started screaming and waving his arms at them. The keepers said it was almost as if he could talk. As it turned out, several young bonobos had entered the dry moat but were unable to get out. The keepers provided a chain as ladder. With human assistance all the bonobos got out except for the smallest one, who had to be pulled up by Kakowet himself.

This anecdote matches my own moat story, which concerns the same enclosure a decade later. By this time, the zoo had wisely decided that water was not needed in the moat (apes cannot swim). A chain permanently hung down into it, and the bonobos visited the moat whenever they wanted. If the dominant male, Vernon, disappeared into the moat, however, Kalind sometimes quickly pulled up the chain. He would then look down at Vernon with an open-mouthed play face while slapping the side of the moat. On several occasions, the only other adult, Loretta, rushed to the scene to rescue her mate by dropping the chain back down and standing guard until he had gotten out.

What both anecdotes tell us is that bonobos seem capable of taking the perspective of someone else. Kakowet seemed to realize that filling the moat while the juveniles were still in it would not be a good idea, even though it would obviously not harm himself. Both Kalind and Loretta seemed to know what purpose the chain served for someone at the bottom of the moat and to act accordingly; the one by teasing, the other by assisting the dependent party. The ability to adopt the viewpoint of someone else is a hotly debated topic in cognitive psychology; it is an advanced capacity, which some believe unique to our species.

This mental capacity has the power of revolutionizing social relationships. To see others as perceiving, feeling, and thinking beings, and to be able to put oneself into their "shoes," makes it possible to sympathize with them, to know what kind of help they need, and . . . to deceive them. Intentional deception requires an awareness of what others know or do not know. An amusing example comes from the zoo nursery where bonobo infants were taken care of. An attendant told little Laura to clean her plate of food. Laura obeyed at once—but not until it was diaper-changing time was it discovered how the food had disappeared so quickly. Laura must have waited till the attendant had her back turned to stuff it into her diapers.[11]

Claudia Jordan provides another example (and similar ones occur daily in a bonobo group). A juvenile male tentatively reached for an apple that had rolled away from his mom's food pile. He stopped, seeming to realize that picking up the fruit would get him into trouble. Instead, he approached his younger sister with an exaggerated play face. She was sitting near her mother in the direction of the apple but had not noticed it. While wrestling with his sister, the male got closer and closer to the apple. Then, with a sudden movement, he grabbed it. His interest in the play vanished, and he went to a corner of the enclosure to quietly enjoy his prize.

It is increasingly believed that perspective-taking and deception are part of an entire "package" of cognitive skills present in apes and humans but perhaps not in other animals. A key finding in support of this new divide—this one not setting us apart from all other animals, but creating a slightly more inclusive elite—is that apes are the only animals other than ourselves to react to a mirror as if they see themselves, not some stranger of their own kind. Evidence for self-recognition comes from an elegant experiment by Gordon Gallup, an American experimental psychologist, in which an ape with a colored dot on his face (put there without him knowing) is placed in front of a mirror. Usually, the ape makes the connection; he inspects and touches the spot on his own face guided by his mirror image. Another indication is the speed with which apes, when first introduced to a mirror, lose interest in their images as social partners. Very soon they stop threatening or socially inviting their images and instead move to activities that are rare or absent in other animals. They inspect body parts of themselves that they normally cannot see, wrinkle their noses, and decorate themselves, placing vegetables or straw on their heads to see how it looks in the mirror.

Bonobos have never been subjected to the official test with the colored mark, but there can be little doubt that they will pass. Both at the Yerkes Regional Primate Research Center and the Antwerp Zoo, the reactions of bonobos to mirrors were videotaped and carefully analyzed. The conclusion was that they reacted like the other apes: self-directed behavior with the help of the mirror, such as picking at the eyes and nose or inspecting the interior of the mouth while staring into the mirror, were common responses. If bonobos indeed join the ranks of self-recognizing animals, this is quite significant because, apart from orangutans, chimpanzees, gorillas, and humans, all other species investigated thus far have failed the critical test. According to some scientists, this means that the ancestor of apes and humans developed a degree of self-awareness unprecedented in the animal kingdom.[12]

What about another traditional yardstick of primate intelligence, such as the use of tools? Wolfgang Köhler described the so-called "aha experience" of chimpanzees when, after staring for a long time at a particular problem, they would suddenly swing into action as if the proverbial lightbulb flashed on in their heads. For example, an ape would be presented with a banana hung from a high point, several boxes, and a long stick. The solution would be to stack the boxes on top of one another, then climb on top of them to knock the reward down with the stick. Because chimpanzees seemed to solve these problems on the basis of planning and foresight, Köhler's experiments presented the first serious challenge to the position of American behaviorists that all problem-solving must be based on trial and error.

Chimpanzees are so good at these tasks that few people were surprised to learn that they use tools in the wild. Jane Goodall described termite-fishing (probing a small stick into a termite mound to harvest insects) and leaf-sponging (using chewed leaves to soak up water from a tree hole). Since tool-making had been declared the single most important trait to set us apart, it was particularly significant that Goodall saw her chimpanzees turning leaves into sponges and modifying twigs by stripping off side shoots before using them as probes. Another uniqueness claim had been debunked. Now we know of many more kinds of tool use in the field, the most spectacular example being nut-cracking: hard nuts are placed on an anvil stone and smashed open with a hammer stone.

If chimpanzees have entered the Stone Age, what about bonobos? In captivity, Jordan saw bonobos scoop up water with red-pepper halves, use a stick to obtain out-of-reach objects or to pole-vault over a moat, wipe their behinds with wood wool, throw with good aim at strangers, sprinkle water on a tennis ball, which they then sucked dry, and so on. Another striking example is Kanzi, a language-trained bonobo, who was presented with a problem that could only be solved with a sharp tool, such as a knife or sharp-edged stone. Kanzi was not given a knife: he received flakes struck off a cobble with a hammer stone by a skilled archaeologist, Nick Toth. Toth was interested in the mental capacities that allowed early humans to develop stone implements, such as the Oldowan tools dating back an estimated two and a half million years.

Kanzi quickly grasped the utility of sharp flakes: he used them to cut a string allowing him access to a baited box. He became discriminating, testing flakes with his lips before taking them to the box, rejecting dull ones. After a while, he tried the flake-making himself, striking two rocks together. According to Sue Savage-Rumbaugh, who raised Kanzi and describes this experiment, he was not very good at creating flakes this way. One day, he invented a better technique, which exasperated the archaeologist, since it was not the way flakes were supposed to be produced. He threw a rock with full force onto a hard surface so that it shattered, producing a whole shower of flakes. When experimenters covered the floor with soft material to thwart this technique, Kanzi simply pulled back the carpet. He did eventually develop a flake-making technique by striking rocks together, and although primitive by Oldowan standards, his tools worked well for the problem at hand.

Even if bonobos show considerable tool-using skills in captivity, including the extraction of honey with sticks from artificial termite mounds, in the natural habitat they have thus far never been seen to probe for insects, sponge water, or crack nuts with stones. If we do not count the dragging of saplings during

(continued on page 41)

INTERVIEW WITH SUE SAVAGE-RUMBAUGH

Sue Savage-Rumbaugh (SSR) and her husband, Duane Rumbaugh, run the Language Research Center of Georgia State University in Atlanta. They work with a number of different primates, but their uncontested linguistic star is Kanzi, a 15-year-old bonobo. The following interview, conducted in February 1996, focuses on the temperament and sociality of their subjects.

FDW: Would you say that you study language or intelligence, or is there no difference?

SSR: There is a difference, because we have apes who have no linguistic abilities in the human sense, but who do quite well on cognitive tasks such as solving a maze problem. Language skills can help elaborate and refine cognitive skills, though, because you can tell an ape who is language-trained something that he does not know. This can put a cognitive task on a whole different plane.

For example, we have a computer game in which apes put three puzzle pieces together to make different portraits. After having learned this, they get four pieces presented on the screen, and the fourth piece is from a different portrait. When we first did this with Kanzi, he would take the piece of a bunny face and put it together with a piece of my face. He kept trying, but of course it wouldn't fit. Since he understands spoken language so well, I could say to him: "Kanzi, we're not making the bunny, put Sue's face together." As soon as he heard this, he stopped making the bunny, and stuck to the pieces of my face. So, the instructions had an immediate effect.

FDW: Since you have worked with both chimpanzees and bonobos, can you say something about species differences in temperament?

SSR: Bonobos are much more group-oriented, they want to sit together and coordinate their activities. When the bonobos are together and we want to separate them, we always have to (laughs) . . . discuss this. The same is not true of chimpanzees: they are more independent-minded.

I don't think chimpanzees are necessarily less concerned about each other, but I do feel they are less aware. If chimpanzees are aware of another's situation, they can be equally protective and caring, but bonobos are more constantly checking on you. And this is not only when you might be in pain or trouble; it applies equally to situations where they may wish to deceive you. They may be planning a trick on you. They are

Sue Savage-Rumbaugh with Panbanisha, an adolescent female bonobo, in the wooded outdoor laboratory around the Language Research Center, in Atlanta. The investigator holds up a panel with symbols, or lexigrams, known as the Yerkish language. Each lexigram stands for a single word, including verbs, nouns, and adjectives. Panbanisha indicates what she wants to do, eat, or play with by pointing at the appropriate lexigrams, and the investigator asks her questions in the same manner.

much more attuned to what you are thinking and what got you to think that way. The social domain is the absolute focus of their life.

FdW: What kind of tricks do they play on you?

SSR: They will send you on an errand. They ask you to look at something else, or to fetch some juice. But the whole purpose is that they have seen that you left a door unlocked, or that you left something behind that they want. They will act as if they haven't noticed, but as soon as you turn your back, they will do the thing that they aren't allowed to do. This is something we need to be constantly aware of.

FDW: You raised Panbanisha [a bonobo] together with Panzee [a chimpanzee]. How do they compare?

SSR: Like the bonobos, Panzee learned symbols just from watching people. She was more than six months delayed in her acquisition of symbols compared to Panbanisha, learned fewer of them, and used them in a more restricted fashion. At present, it is hard to compare them [both apes were then 8.5 years old], because we were unable to continue working with Panzee. So, Panbanisha's symbolic communication now goes far beyond that of Panzee. Who knows, Panzee might have caught up with her.

I should add that on tasks such as puzzle construction, tool use, mazes, and so on, Panzee has always been ahead of Panbanisha. In anything outside the domain of social communication, involving object manipulation or spatial orientation, the chimpanzee was reliably ahead. Whereas in communicatory and perceptual abilities, such as combining television images with the narration, the bonobo was always advanced. The two species may have different cognitive strengths.

FDW: Do Panbanisha and Panzee still get along?

SSR: We keep Panzee now with the other chimpanzees, but I often take Panbanisha to visit her. Panzee is clearly bigger and stronger, and she dominates Panbanisha, but Panbanisha accepts that. They play together, and have a wonderful time.

The only serious problems we had were between Panzee and Kanzi. When Panzee got older, she began to throw things at Kanzi, trying to challenge and needle him. Bonobo females don't do this to bonobo males; when there is a problem, they just turn around and present to them. Panzee did not react this way at all to Kanzi. As a result, he would get upset and scream, and after a lot of back-and-forth between them, they would literally fight. Usually, she would bite him; I don't think he ever bit her.

As soon as I put Panzee in with the chimpanzee males, she quickly learned not to try this with them. It doesn't wash up there. Here with Kanzi she's swaggering all the time, with her hair out, almost male-like, but with the males of her own species she is acting totally different. She is this very nice *troglodytes* female; she fits right in.

FDW: Kanzi looks bigger and stronger than Panzee, though, and he has sharp canines. Would you say he was inhibited?

SSR: Oh, yes, even with Panbanisha he is. For a male bonobo to bite a female is just not done. There can be no doubt about his physical superiority over Panbanisha, but on the rare occasions that the two of them fight, Kanzi always suffers more injuries.

I must add that to say that female bonobos dominate males, as some people do, may not be the right way of putting it. It is true that female bonobos can claim feeding priority—I have seen this with my own eyes at Wamba—but they do not chase the males. It is almost as if the feeding arrangement is agreed-upon. I rather see the relation between the sexes in terms of roles. Each individual has its own role in society, and males and females simply play different roles.

FDW: Do you know good examples of empathy among bonobos?

SSR: At night, we give all the bonobos blankets so that they can build nests, and normally they do this together. But lately I have been doing daily work with Panbanisha. So, our staff may be taking care of the others, giving them their meal, and locking them up. I then take care of Panbanisha later on.

I might give her things, such as raisins or extra milk, that the others did not get. Some of them have dietary restrictions. Panbanisha might point out her favorite foods on the keyboard. As I'd bring it to her, the others would see me walk by, and vocalize. They want the same things, of course. Panbanisha seems to realize this. So, she may ask for juice, and when I arrive with it she simply gestures to the others. And I ask: "Do you want me to give this to Kanzi, or Tamuli?" She then waves an arm in their direction, and vocalizes at Kanzi or Tamuli, who respond with their own calls. They then sit next to Panbanisha's cage, waiting for the food.

I have the impression that Panbanisha wants me to bring the others the same goodies that she is getting.

There exists no obvious connection between the shape and color of the lexigrams learned by Kanzi and Panbanisha and what they represent for the apes.

display or the weaving of branches into a nest, wild bonobos simply do not seem to use any tools at all! There are two ways of looking at this puzzling finding. One is that the propensity for tool use is low in bonobos: the cognitive basis is there, but these apes are little inclined to make use of it. The second possibility is that the bonobo populations studied can obtain all the food they want without tools: there simply is no need for them. If so, we need to find populations surrounded by food sources of high value that can be harvested only by means of implements. If there were still to be no tool use, this would strengthen the first hypothesis.

The absence of tool use in wild bonobos is perhaps best put into perspective by considering an exciting new discovery in orangutans`. These hairy, red apes of Asia are widely reputed in the zoo world as the best tool users among nonhuman primates, as well as the most incredible escape artists. Orangutans are slower and more deliberate in their use of instruments than chimpanzees, but I would bet without hesitation on the orangutan if representatives of all four ape species were presented with the same tool task. It was long believed, however, that wild orangutans did very little with tools. They were known to scratch their behinds with sticks and hold leaves over their heads during a downpour, but there was nothing comparable with the sophisticated tool technology of wild chimpanzees. Until, that is, field primatologists recently encountered orangutans in a Sumatran peat swamp who demonstrated the variety of tool use and the skillfulness we knew this intelligent ape capable of.[13]

Orangutans have been studied for a long time—longer than bonobos—so it would seem prudent to withhold judgment on the bonobo's tool-use performance in the wild until a larger number of populations are known. There exists so much variability in this regard that we now increasingly use the term *culture* in primatology. The reason is that, as in human culture, the habits and innovations of a group seem to be transmitted nongenetically to the next generation. For instance, all the chimpanzees in one community will crack open nuts with stones, and all the youngsters in this community can be seen developing this skill, whereas in another community—which has nut trees and tools available—no one seems even to realize that nuts can be eaten. Flexible behavioral traits transmitted through learning can be regarded as traditions; a set of group-specific traditions can be regarded as a culture.[14]

With regards to cultural variation, the bonobo is understudied. Only one tradition is well documented, but it concerns captive bonobos. In the San Diego colony, it is not unusual for one individual to approach another and slap her own chest with her hand, or clap her hands a few times in front of the other's face, before proceeding to groom. Sometimes the other claps as well, but it is mostly the groomer who claps hands or feet together during concentrated grooming. Transmission of the pattern has taken place since my initial observations: the same gestures are now seen in several individuals introduced later on. Hand-clapping, which appears to express enthusiasm for the grooming contact, has never been reported for any other bonobos: San Diego is the only place in the world where one can actually *hear* bonobos groom.

KANZI

We are a "lateralized" species: the two halves of our brains specialize in quite different functions. In the same way that we are overwhelmingly right-handed when using tools, we rely heavily on the left hemisphere for the perception and production of language. These two biases are related: the right hand is controlled by the left side of the brain. It was initially thought that the connection between language, tool use, and brain asymmetry existed only in our species, but now there is growing evidence for lateralization in apes, suggesting that the connection emerged before language capacities had fully evolved.

Apes favor the left limb for certain tasks (a mother preferentially cradles an infant with her left arm), while selecting the right limb for others (locomotion is often initiated with the right hand). When comparing data on the bonobos at the Yerkes Primate Center and the San Diego Zoo, William Hopkins, an American expert on brain lateralization, and I were excited to discover that handedness extends to gesticulation. Bonobos wave, beg, wrist-shake, or make threatening gestures predominantly with their right hands. This is the first evidence in a close relative of ours that a communicatory capacity other than language may be associated with the left side of the brain. The similarity in brain specialization hints at a shared evolutionary history between gesturing and language.[15]

Is language itself within the bonobo's capacity? Most readers will have heard of Washoe (a chimpanzee), Koko (a gorilla), Chantek (an orangutan), or Kanzi (a bonobo), who communicate in American Sign Language or point at visual icons to request food. Opinions about the complexity of these tasks vary. On the conservative end of the spectrum, we find those who think signing apes are about as interesting as bears on unicycles. Often a comparison is drawn with "Clever Hans," a nineteenth-century horse believed capable of counting. Hans lost his ability, however, as soon as he was tested away from people who could provide unintentional clues. Simply put, the skeptics believe that ape language is in the eye of the beholder.

On the other end of the spectrum, we find scientists who sincerely believe that symbol-using apes stand at the brink of human language, that there is continuity between them and us even in this much-lauded realm. The question is not really whether these animals have language or not; this phrases it too much as an all-or-nothing phenomenon. Language involves a variety of capacities; no one expects apes to demonstrate all of them. Rather, the question is whether they possess some of the basic prerequisites for language. When an ape signs "play outside" or "tickle me," does he or she rely on brain structures and mental capacities equivalent to those that we use when we formulate a request? The jury is still out on this one.

Personally, I must admit to mixed feelings about ape language research. On the one hand, I see it as a thoroughly anthropocentric enterprise. A communication system for which evolution has specifically hardwired us (and perhaps only us: our brains are three times larger than the average ape brain) is being imposed upon another creature to see how far it can go. There is something inherently unfair about judging them on our terms. Might we not learn more about them by scrutinizing their own communication systems, such as their hand gestures or vocalizations? On the other hand, the apes in these studies are so well attuned to people, so willing to interact, so used to the way we relate to our surroundings, that all sorts of questions can be addressed that are impossible to answer with apes who view us as strangers with strange habits. As such, this kind of research opens an important window on the ape mind. It allows us to explain to them what we want and to ask them how they perceive things. Kanzi's flint-making is an example of an experiment that might not have worked with an untrained subject.[16]

Kanzi's abilities go well beyond success on this relatively simple tool task, however. As a young infant, he picked up a sizable vocabulary of keyboard symbols without any rewards. He did so simply by sitting in on the training sessions of his adoptive mother, Matata, who learned very little during the same period. According to Sue Savage-Rumbaugh, Kanzi makes rudimentary sentences of two or three utterances, the ordering of which shows some regularity. This could be interpreted as a grammarlike structure, a claim hotly contested by linguists for whom grammar is the touchstone of language.

Even more than his utterances, however, Kanzi's understanding of spoken English is truly impressive. Many animals are able to guess what we mean when we talk to them. They probably do so by paying attention to our tone of voice, gaze direction, and the entire context in which the interaction occurs. Kanzi, however, listens through earphones to a battery of words spoken by a person in another room, thus preventing unintentional cluing. Without hesitation he selects the corresponding picture from among a pile of pictures. He hears "melon" and picks the melon picture; he hears "cat" and picks the cat

Kanzi, the world's most famous bonobo, has been featured in numerous newspaper articles and television documentaries. The presumed gap in intelligence and consciousness between humans and apes seems rather narrow to anyone meeting this anthropoid genius up close.

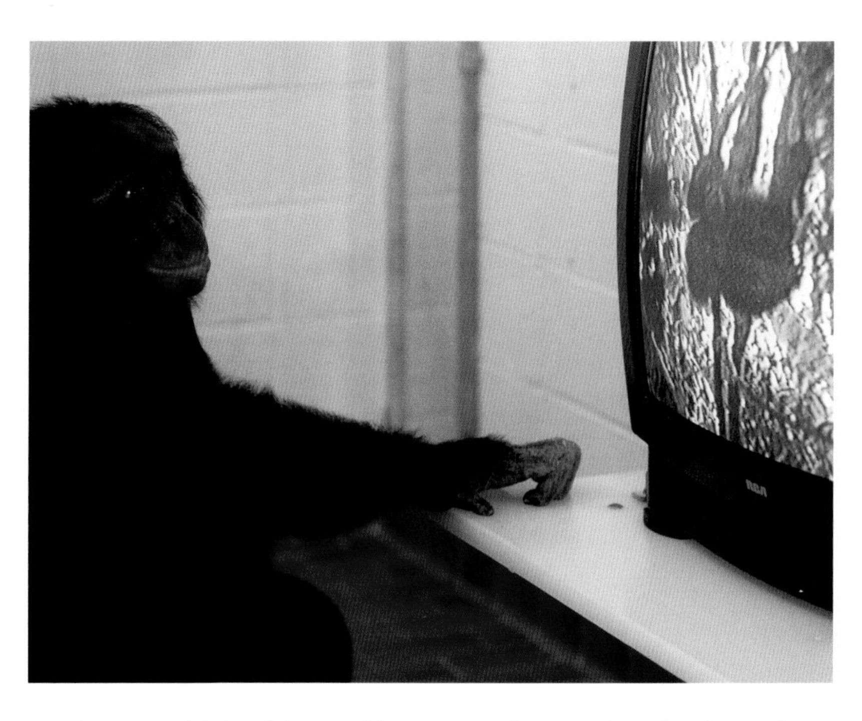

Apes love to watch TV, and they are able to recognize the action that takes place on the screen. Does Kanzi, following the movements of a tree-dwelling ape, have any idea that this is the life he never lived?

picture. His comprehension skills go even further: based on what he hears, he is able to connect different objects. For example, if Savage-Rumbaugh tells him, "Give the dog a shot," he picks up a hypodermic syringe from among many objects on the floor, tears off the cap, and injects his stuffed toy dog.

Savage-Rumbaugh does not believe, as some linguists do, that language comprehension is a piece of cake compared to actual language production: "From my perspective, comprehension is the essence of language, and is far more difficult to explain and to achieve than production," she observes. "Comprehension demands an active intellectual process of listening to another party while trying to figure out, from a short burst of sounds, the other's meaning and intent—both of which are always imperfectly conveyed. Production, by contrast, is simple. . . . We don't have to figure out 'what it is we mean,' only how to say it."[17]

Whatever the implications for language, it is impossible not to be struck by Kanzi's obvious intelligence. For me, watching him raises more questions about bonobos. What do these apes normally—in the forest, in the social group, with respect to tools—use all this brain power for? I wonder about the natural function. It is a bit like Little Foot's notorious toe: if it is there, you would think it served a purpose. Kanzi's wild counterparts wear no earphones, yet they may well listen more carefully to one another than we dare assume. It could well be that they make connections between elements or events in their environment that none of us are yet aware of.

THE UPRIGHT APE

The half-upright posture of this aging male (about 40 years old) shows that little evolutionary change would have been required for an ancestor with a bonobo-like anatomy to develop a bipedal gait. Some scientists claim that the bonobo is the best model of the ancestor common to apes and humans, but others consider it a relatively specialized form.

Two typical locomotion patterns of the bonobo are climbing (opposite, a male) and knuckle-walking (above, a female and her infant). In knuckle-walking, the weight is carried by friction pads on the middle phalanges, with the fingers curled inward. Such terrestrial locomotion counters the evolutionary pressure to shorten the fingers: long fingers are advantageous in the trees. Only the African apes evolved knuckle-walking; monkeys walk on the palms of their hands.

Bonobos are excellent bipeds. True to some theories on the origins of the human upright posture, they often walk on two legs when carrying food; here, a captive female carries ginger leaves (opposite), and a male in the forest carries the sugar-cane that researchers have provided (right).

Two individuals show off their elegant limbs. The bonobo's legs are remarkably long in comparison to those of other apes. Notice too the frontal orientation of the female's genitals (above) and the large size of the male's testicles (opposite).

CHAPTER 3

IN THE HEART OF AFRICA

Because the Mercator projection grossly exaggerates the dimensions of the northern hemisphere, Zaire appears on our maps as a medium-sized country. Yet it is huge: about the size of the United States east of the Mississippi, or all of western Europe without the Scandinavian countries. It has only around 40 million people, though, 40 percent of whom live in urban areas. Nearly 80 percent of Zaire is covered by forest. Formerly known as the Belgian Congo, most of the country still fits the image evoked by its colonial name: the dark, unsubdued heart of Africa.

The Cuvette Centrale in northern Zaire is part of the second-largest solid stretch of rain forest in the world. All one sees from a plane is a ceaseless canopy of tree crowns, without trails or clearings, covering the flat basin of the mighty Zaire River. The forest called home by the bonobo is surpassed in surface area only by the jungles of the Amazon. One might think that the vast dimensions of the natural habitat, together with its inaccessibility, guarantee the ape's survival. Bonobo experts desperately wish this were true—and cling to the hope of surviving populations in unexplored regions—yet the picture is bleak. Fieldwork on the species, which has naturally become mixed with con-

The many small rivers that crisscross the bonobo's home range in Zaire present major obstacles to dispersal, since apes do not swim.

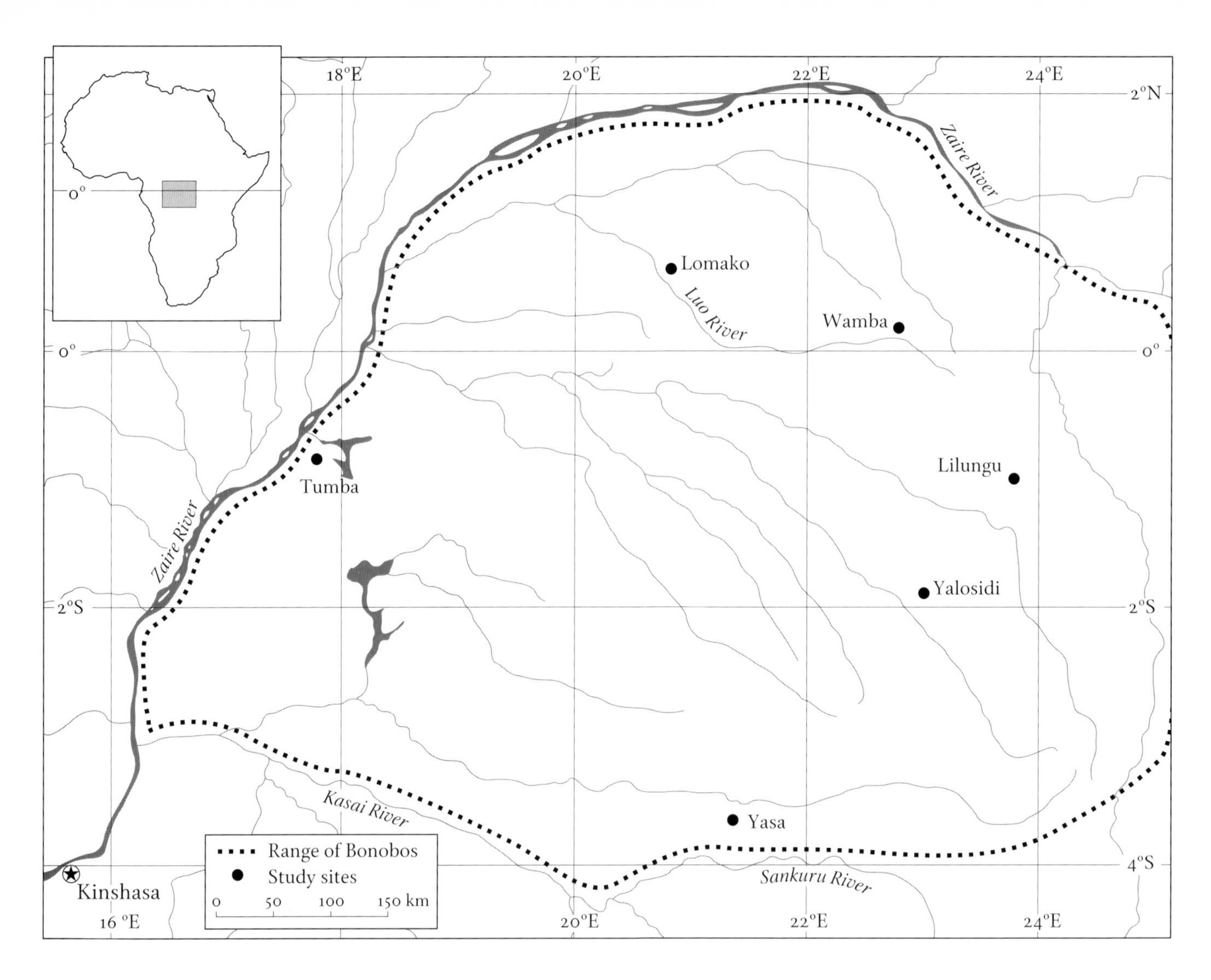

In the 1920s, it was still widely believed that no apes lived south of the Zaire River. In 1927, however, the Tervuren Museum in Belgium received an ape skull from a town not far south of Lomako—the specimen that led to the discovery of the bonobo two years later. This map of northern Zaire shows the location of six current sites for bonobo research, with Yasa as the southernmost site.

servation efforts, seems in a race against the clock. Actually, the current surge in scientific interest may well turn out to be the bonobo's lifesaver: there is hardly a more effective way to protect a tropical species than through the presence of researchers in the same area.

The first decade following Zaire's independence from Belgium in 1960 was marked by bloody strife among its hundreds of tribes. For this reason, foreign scientists avoided the country until well into the 1970s. While long-term projects on chimpanzees, orangutans, and gorillas were being set up in Tanzania, Borneo, and Rwanda in a surge of fieldwork on apes, the only region to harbor bonobos was left out. Few scientists realized in those days that bonobos were not just a variety of chimpanzee but a unique species, worthy of the same amount of attention. The fact that bonobos are rare in captivity also did not help: most scientists had never seen one.

In the wild, it is hard to find bonobos for an entirely different reason: they are afraid of people. The very first field project, set up in 1972 by an American anthropologist, Arthur Horn, is now almost forgotten, having resulted in only six hours of observation of bonobos over a two-year period. In 1974, however, as already noted, the Badrians and Takayoshi Kano independently initiated fieldwork. They were more successful in getting close to the apes, and both field sites are still operative today. Wamba, where Kano settled, is located directly on the equator (latitude 0°10′ N, longitude 22°34′ E), whereas Lomako, where the Badrians went, is to the west and north from there (latitude 0°51′ N, longitude 21°05′ E). Both sites are in the northern part of the bonobo range, which is estimated to be bounded to the north and west by the Zaire River, to the east by the Lomami River, and to the south by the Kasai and Sankuru Rivers. The area of potential distribution is over 800,000 km^2. According to Kano, however, the range in which bonobos can actually be found is probably closer to 200,000 km^2, or an area the size of Great Britain.

The following descriptions of the life of bonobos in the forest summarizes information gathered by others who have made regular expeditions—under very difficult and sometimes dangerous circumstances—to Central Africa. I myself am not a fieldworker. Without the hard-won evidence from Wamba, Lomako, and the other sites, no coherent picture of bonobo natural history would be possible. It is to be hoped that the apes will survive in large enough numbers for this important research to continue.

INTERVIEW WITH TAKAYOSHI KANO AND SUEHISA KURODA

Kano is director of the bonobo project at Wamba, Zaire. Kuroda is his long-time collaborator. Both are affiliated with Kyoto University. The interview took place in November 1994.

FDW: Dr. Kano, can you describe your very first encounter with bonobos?

KANO: In 1974, I encountered ten individuals in the forest of Yalosidi. I was impressed how different their calls were from chimpanzees: they had such high voices! I also felt that they were more arboreal than chimpanzees. Each time I have encountered chimpanzees (in Tanzania), they would drop straight down out of a tree and flee running over the ground. The bonobos, however, fled for more than one hundred meters through the canopy before they descended to the forest floor.

FDW: What would you say are the main differences in social organization between bonobos and chimpanzees?

KANO: In the bonobo we see (1) long-term bonding between mother and son, (2) the use of sexual behavior as a social tool, and (3) a female-centric organization. We have now six confirmed mother-son dyads, who are still together even though the sons are fully grown. These combinations travel almost always together, in the same party.
KURODA: The long dependence of the son may be caused by the slow growth of the bonobo infant, which seems slower than in the chimpanzee. For example, even after one year of age, bonobo infants do not walk or climb much, and are very slow. The mothers keep them near. They start to play with others at about 1.5 years, which is much later than in the chimpanzee. During this period, mothers are very attentive. The two may develop a very intense communication because an infant unable to reach certain objects will call the mother, and the mother then comes over to help.*

FDW: I can see how this builds a strong attachment, but wouldn't this apply equally to daughters and sons? What happens with the daughters?
KURODA: Female juveniles gradually loosen their tie with the mother and travel further away from her than do her sons. This is quite distinct, sometimes we see young females who stay alone, or are peripheral to a party, even if they are only five or six years of age. Sons, however, are always near their mother. In fact, in talking with people, I often hear the same comment from human mothers—that is, they say sons are more affectionate and attached than daughters.

FDW: Are female bonobos dominant, equal, or subordinate to males?
KANO: They are almost co-dominant. Males seem slightly dominant because the alpha male [the highest ranking adult male] in my main study group is the only male who can threaten or charge the two top-ranking females.

FDW: Do these females avoid him?
KANO: No, this is very interesting: when the alpha male charges at the alpha female, she usually completely ignores him. [*Both scientists laugh.*]

---

*The delayed growth of the species and its dwarfish size during infancy and early juvenescence is reflected in Robert Yerkes's underestimation of Prince Chim's age, as well as the adoption of the French word for "peanut" as a name for San Diego's breeding male (see chapter 1).

FDW: What happens with food? Can the alpha female take it away from the alpha male?

KANO: Yes, of course. She is dominant when it comes to food.

KURODA: I believe that social dominance cannot explain all of the encounters. When they are competing over food, the most impressive thing is that the females behave exactly as they like, whereas the males cannot do so. Even if males are physically stronger, they do not wish to fight against females. The male might be able to win, but all he can do at such times is run away.

Females are good at ignoring beggars: "Why are you bothering me, can't you see I'm eating?" But when the possessor is a male, and he is approached by a female, his self-confidence evaporates.

FDW: Which species do you think is smarter: bonobo or chimpanzee?

KURODA: The bonobo, of course! [*We laugh about such undisguised bonobocentrism.*] There are so many kinds of intelligence. If you compare the ability to use tools or to use partners for strategic purposes, bonobos are not particularly smart. But if we look at intimate social relationships, their cognition is highly developed due to their long dependency as infants. In the domain of attachment, affection, and the avoidance of conflict, they are very intelligent. For example, chimpanzees are unable to develop peaceful relationships with other groups. Their social organization focuses on how to gain advantage and how to fight other groups. The peacefulness of bonobos rests on their ability to recognize the value of social relationships.

FDW: Why don't they use tools in the wild?

KANO: They do not need them. There are many termites, but they do not eat them.

KURODA: There are alternative sources of protein. For example, there are occasionally lots of caterpillars in the forest. I'm sure bonobos will use tools whenever they need to.

FDW: What do you think of the future of bonobos in Zaire?

KURODA: The population is decreasing rapidly because of hunting for meat. Hunting pressure increased dramatically after the civil war, in 1991, because the unrest destroyed food supplies throughout the country. People were hungry, forgot about the old traditions (such a taboos on killing bonobos), and ignored the law. We need to be very worried about

Takayoshi Kano, who teaches at Kyoto University, is the director of the longest-running studies of bonobos in the wild. He arrived in Wamba on bicycle in 1973 and has managed to keep the site operative for over two decades, despite primitive conditions and political instability. Without his pioneering work, science would still be very much in the dark about bonobo social life.

the bonobo's future. Five years ago we thought the total population might be close to 10,000, but now it may be less.

Yalosidi forest is a typical example. In the 1970s, Kano found a very high population density there. When our colleague, Gen'ichi Idani, went to Yalosidi about ten years later, the bonobos had virtually disappeared!

FDW: Could it be that there exist populations that we don't know about, or that bonobos are so shy in some places that we never get to see them?

KANO: Yes, they are often shy, but even then one would find nests, because bonobos build a nest every night. You can tell if there are apes in the forest by the nests. I think they are very endangered. Twenty years ago, hunting was rare in much of their range, but now bonobos are increasingly eaten by people. Even Wamba is affected.

KURODA: Zaire is a large country, and in some areas the human population is extremely low. I still have some hope that there may be undisturbed forests with large bonobo populations. It is very urgent that we find such places . . . if they exist.

THE WAMBA APPROACH

Takayoshi Kano belongs to the world-renowned school of primatology at the University of Kyoto, which developed after World War II out of behavioral studies on free-ranging horses and Japan's native macaques, the so-called snow monkeys. The perspective of the Kyoto school is somewhat different from that of traditional Western science. Instead of focusing on competition and the Darwinian struggle for life, Japanese primatology emphasizes the social fabric. One sometimes hears that Japanese business takes a long-term view rather than the quarterly view common in the West; in the same way, Japanese primatologists invest in the collection of data that often can be expected to bear fruit only years later. Long-term records and recognition of hundreds of individuals are a staple of their research, along with limited provisioning of food in order to attract the animals. Through patient observation of monkey populations for generations, Japanese scientists were the first to demonstrate lifelong kinship bonds. The founder of the Kyoto school, Kinji Imanishi, further advocated identification with animal subjects: he argued that a certain level of subjectivity is essential for a full understanding of behavior.

The influence of Japanese primatology has been such that its approach now sounds commonplace. Identifying individuals by name or number, for example, was long resisted in the West (opponents argued that it made animals too humanlike), but is now routine. Many primatologists now follow primates over their entire life spans and feel that empathy by the human observer is both inevitable and desirable. The possibility of animal culture, first suggested for Japanese macaques, is also gaining acceptance. Meanwhile, an entire new generation of Japanese primatologists has been educated in the statistical methods and evolutionary perspective of Western biology. Today, Western and Eastern approaches have merged.

Observations by Kano and his co-workers demonstrate many distinctions between bonobo and chimpanzee societies, as well as one fundamental similarity. Both species live in so-called *fission-fusion* societies—that is, the apes travel in small "parties" of a few individuals at a time, the composition of which changes from hour to hour and from day to day. All associations, except the one between mother and dependent offspring, are of a temporary character. Initially, this flexibility baffled investigators and made them wonder whether these apes knew stable social groupings at all. After years of carefully documenting the composition of chimpanzee parties in Tanzania, Toshisada Nishida, a close colleague of Kano's, was the first to crack the puzzle. He reported that chimpanzees form large unit-groups, or communities: all members of a particular community mix freely in ever-changing parties, but members of different communities never gather.

Both species are male-philopatric—that is, males stay in their natal groups, whereas females disperse to neighboring groups. As a result, the senior males of a chimpanzee or bonobo group have known all the junior males since birth, and all the junior males have grown up together. Moreover, males are often related to one another, and there are close bonds between maternal brothers. Females, on the other hand, transfer to an unfamiliar and often hostile group, where they may know no one or only a few female migrants from their own community.

To discover all this in forest-dwelling apes is more complex than it appears. The fieldworker may notice that a particular female has disappeared, but this could be for a multitude of reasons. In a large forest, with traveling parties constantly changing composition, an individual may go unnoticed for a long time before it is considered missing. Only if this female then shows up in a neighboring group—which means that one needs to have habituated and studied that particular group as well—will one be able to deduce what happened. Or take kinship: it sounds simple to say that brothers form close bonds, but one can only know for sure which adult males are brothers if one has seen them grow up together attached to the same female. Siblings often resemble each other, but no self-respecting fieldworker would trust such subjective criteria. Since apes mature slowly, it takes at least a decade of dedicated observation before reliable statements about the kinship structure of a community can be made.

While other studies have focused on how bonobos exploit the resources around them, the party size in which they travel, and the way they locomote through the trees, the Japanese team has collected this sort of information chiefly as background for the much more ambitious task of tracking social relationships. The emphasis is not on the species's ecology but on its social behavior. Not that these two subjects are antithetical. It is widely assumed that social life is adapted to ecological conditions: ideally, both are studied. Methodology, however, follows logically from theoretical emphasis. Ecologists frown on Kano's provisioning of sugarcane. From their perspective, the Wamba bonobos are compromised study subjects: provisioning is bound to affect food selection and ranging patterns. The Kyoto team might never have been able to accomplish what they have, however, without a reliable technique to get the bonobos used to their presence. They, in turn, wonder about the usefulness of brief studies of poorly habituated apes. Did the investigators know the social context of what they saw? Did they manage to see a representative set of individuals?[1]

Because only the Wamba scientists have thus far made the sustained effort necessary to know the life histories of individual apes, the information from their site, described in Kano's book *The Last Ape* and in a stream of articles by him and his colleagues, is absolutely invaluable.

PARTIES IN THE FOREST

Apart from watching bonobos at the feeding site, the Wamba investigators follow them into the forest, of which there are three kinds. Swamp forest near the rivers includes relatively low trees, which are supported by prop roots, or by leaning against each other, because of the loose, muddy soil. Primary forest grows on a firmer foundation and is darker because of the dense, overlapping tree crowns and the much taller trees, including giants of up to 50 meters high. The lack of light makes for relatively sparse undergrowth. Finally, there is the secondary forest resulting from clear-cutting. After the people have left the area, vegetation soon covers their traces, but even when full-sized trees have returned, tree density remains below that of the primary forest. Thick patches of herbs grow at places where sunlight reaches the forest floor. The bonobos visit all three kinds of habitat, but they prefer primary forest, which has the richest botanical variety.

At Wamba and elsewhere, bonobos eat a large assortment of plant foods, chiefly ripe fruits. Most of these are the size of figs, but some can reach extraordinary dimensions. *Anonidium* fruits occasionally grow to 10 kg, and *Treculia* fruits even to 30 kg, almost the weight of a full-grown bonobo. These colossal fruits are usually consumed after they have dropped to the ground, and many other edible fruits ripen on lower vegetation. Thus, foraging is not limited to the trees, but includes the forest floor. After fruit, the second-most consumed food is pith from plants known among specialists as THV, or terrestrial herbaceous vegetation. As the bonobos travel from place to place, they stop to snack on the leaf petioles and shoots of these herbs. Kano notes of some of these crunchy vegetables, which are also collected by the local people, that they have an "indescribably delicious" taste. This food has a high protein content, which may explain a major difference from the food intake of chimpanzees: bonobos seem less dependent on animal protein.

Chimpanzees eat notable quantities of animal foods, including monkeys, which they hunt and kill. Meat-eating is more sporadic in bonobos, and their relation with monkeys is far removed from that between predator and prey. In Wamba, monkeys have actually been seen to groom and play with bonobos, but Spanish primatologists in the Lilungu region, a new field site, have observed forced interactions between apes and monkeys. The bonobos seemed to regard monkeys as toys. Jorge Sabater-Pi relates how on three separate occasions, after bonobos had captured monkeys, they inspected, hugged, groomed, or mounted them. They also tossed them in the air or swung the poor animals by their tails, sometimes banging their heads. According to the fieldworkers, it looked as if the bonobos wanted to play and became rough when the monkeys failed to cooperate, despite having been groomed by the eager apes. One victim escaped, but two monkey infants died in the process. There was no evidence that they were eaten; indeed, one bonobo carried the dead body around for two days before it disappeared.

This behavior contrasts sharply with that of chimpanzees, who would not have hesitated to tear the monkeys apart and eat them. Perhaps bonobos fill their protein needs largely by munching THV. They do on occasion consume animals, but mostly small ones, starting with insects. In the forest, the droppings of thousands of caterpillars gathering on the huge botuna trees sound like drizzling rain. The caterpillars swarm every year and are voraciously eaten not only by the bonobos but also by people. The bonobos also eat earthworms, which they dig up from the mud. When hen's eggs were offered at the Wamba feeding site, the bonobos ate them or carried them away. Given their reluctance to try unknown foods, this probably means that they collect eggs from bird nests. As for vertebrate prey, evidence now has accumulated from various sites for the consumption of reptiles, shrews, flying squirrels, duikers (small forest antelopes),[2] and perhaps small fish or shrimp. It should be stressed, though, that on the basis of one thousand fecal samples, Wamba investigators estimate that animal foods comprise only about 1 percent of the bonobo's diet.

Bonobos forage in small groups, or parties, of ever-changing composition. How their party organization differs from that of chimpanzees, who follow the same fission-fusion pattern, is intensely debated. Since party size depends on food distribution—parties grow larger if individuals can feed together, such as in large fruiting trees—provisioning by investigators very likely has an impact. This makes it hard to compare unprovisioned bonobos with provisioned chimpanzees, and vice versa. Recently, two studies got around this problem. Takeshi Furuichi and Hiroshi Ihobe compared the Wamba bonobos with a provisioned group of chimpanzees in the Mahale Mountains in Tanzania. Frances White and Colin Chapman made a similar comparison between the Lomako bonobos and an unprovisioned group of chimpanzees in Kibale National Park, in Uganda. Now we have two detailed comparisons, one in which both species receive extra food and the other in which both do without.

White and Chapman found that chimpanzee females tend to move apart whenever they find themselves near one another, whereas in bonobos it is the males who stay away from one another. All other sex combinations show tolerance and cohesion—that is, individuals remain together once they have gotten close. This suggests that the chief difference is between male bonding and female aversion in the chimpanzee and female bonding and male aversion in the bonobo.

The data of Furuichi and Ihobe from Wamba, however, provide a slightly different picture. They, too, found high affiliation among female bonobos, but further report that males of this species travel together and groom one another just as much as male chimpanzees in Mahale do. Whether this is characteristic of the species remains to be settled, but it does suggest that bonobo males have a potential for bonding and mutual tolerance. The same can be said, by the way, for female chimpanzees. Even though female bonding is not pronounced in East African populations—where females mostly travel on their own with their dependent offspring—such bonding has been reported for some populations in West Africa, as well as for captive chimpanzee colonies, in which females commonly groom one another, share food, and protect members of their own sex against male aggression.[3]

What this boils down to is that the two species emphasize different intrasexual relations, with female-female contact being more important in the bonobo and male-male contact more important in the chimpanzee. The difference is only a matter of degree, however: there is great flexibility within each species. The one consistent species difference is a closer relationship between the sexes in bonobos. At Wamba, three-quarters of the traveling parties are mixed: they include adults of both sexes and mothers with offspring. The figure for Lomako is only slightly lower. At both study sites, grooming (a widely accepted gauge of social ties) is most common between the sexes, followed by grooming among females, with the least amount of grooming among males.[4]

In the chimpanzee, males will travel with a sexually receptive female, but since females are only occasionally in this attractive state, the sexes rarely seek each other's company. Female bonobos, on the other hand, sport the pink swellings that signal receptivity for extended periods of time; parties virtually always include at least one swollen female. Perhaps as a result, the sexes are more often together in mixed parties. Furthermore, male bonobos—including fully grown ones—follow their mothers around through the forest, which makes for even more intersexual association.

Since females of both species migrate to other groups, the only close kinship ties that get a chance to form are those between mother and son and between brothers (fatherhood is unknown to the investigators and almost certainly to the apes as well). Chimpanzee brothers tend to associate and support each other in fights, such as the alliance between Faben and Figan that allowed Figan to conquer the top spot in the Gombe community. In contrast, the mother-son bond in chimpanzees, although clearly in evidence, is only minimally developed. Bonobos show the opposite pattern: the focus of male kinship bonding has shifted from siblings towards the mother. That the mother-son tie also has implications for the male rank order makes the parallels and contrasts with chimpanzees all the more intriguing. The role of the mother is so all-important that Kano has called mothers the "core" of bonobo society.

Bonobos seem more gregarious than chimpanzees: they rarely move alone, and the average party at Wamba is close to twenty. This is several times larger than the typical party size of chimpanzees. It has been suggested that provi-

OPPOSITE: A bonobo male in the darkness of the forest floor.

sioning may have something to do with this, since Wamba's party size is also large compared to Lomako, where the average is about seven. This cannot be the whole story, however, as some of the early data from Wamba stemmed from unprovisioned groups. Suehisa Kuroda, who explored bonobo populations away from Wamba's main site, felt that large party sizes were common in the region. The explanation may be ecological rather than methodological: Wamba is richer in THV than Lomako, and the predominant fruit tree at Wamba is not found in Lomako.

In terms of community size, the two sites are similar, however.[5] Bonobo communities—of which parties are only temporary manifestations—range in size from 25 to 75, but there are estimates of communities of up to 120 individuals. Chimpanzee communities show a similar variability.

Another sign of the bonobo's gregarious nature is that community members gather at night. Every evening a bonobo takes a few minutes to interweave branches high up in a tree so as to construct a comfortable platform, upon which he or she then spends the night. Barbara Fruth and Gottfried Hohmann found nesting parties in Lomako to be consistently larger than traveling parties. In other words, clusters of nests include more individuals than typically found together in the daytime. We know very little about possible dangers to these apes' lives; perhaps the presence of large predators, such as leopards, explains why they call one another when it gets dark and seek one another's company at a roosting site.[6] The investigators further speculate that information about plentiful food sources may be exchanged at these gatherings.

Fact is that, however much bonobos fission during the day, they tend to fuse at night.

INTERVIEW WITH FRANS LANTING

In July 1996, I interviewed Frans Lanting about his 1992 assignment to photograph the bonobos of Wamba, in Zaire. His rich field background gives him an uncommon perspective on the site and its inhabitants.

FDW: Do you feel that the bonobos were a tough assignment?

LANTING: In the past twenty years I've had the privilege of spending extended periods of time with many kinds of animals in exotic places all around the world, yet photographing bonobos was one of the most exhilarating professional experiences of my career. When I saw them up close in the forest I really had to blink—that's how much they reminded me of ourselves. I related to them in an immediate, emotional fashion,

but conveying this in photos was something else. It is easy enough to recognize their gestures and facial expressions, but these signals only make sense in the wider social context, which is virtually impossible to capture on film. What I ended up photographing was only the tip of the iceberg in terms of bonobo society.

In a tropical forest the light level is so low, with harsh highlights and deep shadows, that it's hard to take good pictures. Add to this the fact that the apes are black, and their eyes dark, and it becomes clear that the photographic challenge was formidable. Bonobos are surprisingly fast and exceptionally hard to follow. To keep up with them in a steaming hot rain forest with thick undergrowth when you are carrying twenty kilos of photo gear is a major enterprise in itself. Furthermore, camera lenses tend to fog up in the extreme humidity of the jungle. In the mornings I often had to wait an hour for my lenses to clear before I could shoot.

Some subjects lend themselves to creativity. In the case of bonobos, a big part of my time was taken up by technical problems—not to mention the physical difficulties of survival in a bush camp in central Zaire. As a result there was little room for the kind of experimentation my work is known for. My main goal was to show how close we are to bonobos and they to us. Occasionally they revealed their unique and typical bonobo ways of doing things, and I hope I have captured some of those moments—such as the one scene that shows them all out in a clearing, standing upright and vocalizing intensely. I later worked with captive bonobos to show close-ups and subtle expressions that are virtually impossible to catch on film in the forest.

FDW: Did you find Wamba accessible?

LANTING: Zaire is one of the few countries in the world where travel has actually gotten more difficult over the past fifty years. If you open a Michelin map of Africa, with Zaire at its heart, and see two locations no farther apart than the width of a thumbnail, you may still be looking at a distance that will take several days of travel, if you can get from one to the other at all. A few years ago a British camera crew tried to reach central Zaire by traveling overland from the east, from Kenya. It took them almost two months!

Support from the National Geographic Society for my expedition allowed me to charter a plane. But even that required such logistics as radioing ahead to a mission near Wamba so that the missionaries had time to hire local people to cut the grass on the runway before our arrival. That's how rare planes are in the Djolu region.

FDW: How are the living standards?

LANTING: Ever since independence, Zaire's economy has been disintegrating. It is almost impossible for people to achieve more than a minimum subsistence level. Wamba used to have coffee plantations. Every now and then a coffee trader with a truck would show up to haul off their product. But now the roads have deteriorated to the point that this is not possible anymore. What little money the local people used to earn is gone. Some have started making their own clothes again. Instead of using fabric, they are going back to wearing grass skirts. When they have no matches, I was told that some people are again making fire by rubbing sticks together.

Perhaps the most telling detail is that they have picked up the old custom of communicating with talking drums. Important news is relayed by means of drum signals that carry several kilometers and are relayed from one place to the next. Each of the investigators at Wamba now has a drum name. Takayoshi Kano's name literally means something like "the man with a strong will."

FDW: What can you tell us about life in the camp?

LANTING: The Japanese researchers are a tough lot. They lead a very basic existence in a compound next to the village of Wamba. They have mud huts just like the local people, and their only luxury—understandable given how important bathing is in Japanese culture—is a little bathhouse that is totally unique in the region. When you return from a hot day in the forest, there really is no better way to restore your internal harmony than by soaking yourself in hot water, with one bucket allotted to each person.

The food is rice, local vegetables, and sometimes meat. Eating three meals a day, as in the camp, is probably a lot more than local people can afford to do, but the people are not starving. After all, they live in the humid tropics, where things grow easily.

I was struck by the close relations between the researchers and the local people. There is a lot of debate in conservation circles about what the best policy is for preserving natural habitats. Perhaps national parks are not the answer in Africa, or not the only answer. What Kano is doing in Zaire is, in the long run, probably one of the most effective methods. No primatologist comes to Wamba without first having learned Lingala, the local language. And since the researchers employ lots of people, all of Wamba's villagers are keenly aware of the value of the bonobos for their economy. Also, by employing people from many families and villages, Kano has extended his political alliances in a way that every primate would understand.

FdW: Do any of your photographs stand out in your mind?

LANTING: I was fascinated by the tensions and connections between people and bonobos—both the threat of a growing human population and the physical proximity of two such closely related species. I saw an opportunity to evoke this when we were following bonobos in the forest and, coming to a road, suddenly heard children who were walking home from school. I positioned myself on the road in the hope that the apes would cross. That's how I captured the image of a family of bonobos on the run with a young generation of humans watching in the background. Apes and humans—that's what my work is all about.

WHO'S THE BOSS?

Students of animal behavior are accustomed to ranking individuals from high to low on a scale of dominance. This is easy to do with male chimpanzees and baboons, as well as with the females of many Old World monkeys, such as macaques and vervet monkeys. Social rankings were discovered in the 1920s in domestic fowl based on the direction of attacks among hens (hence the term *pecking order*). Apart from dominance reflected in the outcome of conflict, many animals possess status displays. These displays function a bit like the stripes on a military uniform; they signal an individual's rank. In chimpanzees, for example, the dominant makes himself look large by raising his hair and standing upright, whereas the subordinate literally grovels in the dust uttering panting grunts.

That bonobos lack such formalized rituals of dominance and submission tells us already how relatively unimportant status must be in their society. This is particularly true of relations among adult females. Status is not wholly absent, but it is so vague that Kano does not wish to speak of "high-ranking"

At Wamba, Kano follows food-provisioning techniques developed by researchers of the snow monkeys in his home country. By growing sugarcane for the bonobos, he enticed them out of the forest and won their trust. Several bonobos are waiting at the feeding site.

females, only of "influential" ones. He claims that females "are respected out of affection, not because their rank is high."[7] Aggression does occur among females: one female may without warning jump on top of another, bite her, and steal her sugarcane. It has even been suggested that female-female fights are the worst possible fights in bonobo society, yet they constitute such a minuscule proportion of aggressive disputes that for all intents and purposes, females can be said to be remarkably tolerant.[8] If there is a female rank order, it is largely based on seniority rather than physical intimidation: older females are generally of higher status than younger ones.

Conversely, the lowest-status females are recent immigrants from other communities. These young females keep a low profile, avoid getting involved in fights, and seldom draw attention to themselves. Upon transfer into their new group, they single out one particular resident female for special attention: they try to groom her and invite her to sexual contact. According to Gen'ichi Idani, who has studied this process, close friendships are established if the resident reciprocates. This contact helps the immigrant become accepted into the close-knit female community. After having produced her first offspring, the young female's position becomes more stable and central until, when she grows older and climbs in status, the cycle is repeated, with young immigrants now seeking a good relationship with her.

Among males the situation is entirely different. As far as we know, males do not move between groups, and dominance seems to matter a great deal to them. There is much more fighting among males than among females. Whereas rank positions near the top, specifically the position of alpha male, tend to be quite clear, mid-ranking and lower positions are not so well defined. Since bonobos do not show elaborate status rituals, the rank order is mostly expressed in the direction of aggressive chases. These encounters, which rarely escalate, often end in a quick conciliatory contact in which two males mount each other or rub their scrotums together standing back-to-back. Mounting may occur several times, with the males changing position, and rump contact is mutual by definition, so that these contacts carry a message of symmetry rather than inequality. The level of tension is quite a bit lower than that among male chimpanzees, who also reconcile after fights, but usually after some delay and with gestures and signals that underline the hierarchy.

The greatest contrast, however, is in the factors that determine a male's rank. In chimpanzees, the key is alliance formation among males. A challenger will try to recruit other males to help him overthrow the established leader, and if successful, the new alpha has an "obligation" towards his allies (for example, he will allow his allies, but not his rivals, to mate with sexually receptive females). Such deal-making has been suggested for wild chimpanzees, and I documented it in detail myself in the large colony at the Arnhem Zoo in the Netherlands. The social maneuvering among the Arnhem males was so complex and strategic that I dubbed it "chimpanzee politics."

If there is such a thing as bonobo politics, it more than likely revolves as much around females as around males. According to Kano, male fights are usually one-sided and over quickly, whereas female fights, although rare, may create great confusion, since other females are drawn into the fray. It is possible, though, that the confusion is mostly in the eyes of the human observer:

the animal actors themselves may be anything but confused. If one were to analyze what happens at such moments carefully (for example, with the aid of video), one might uncover an intricate network of alliances. I suspect that female bonobos establish an order among themselves that requires little reinforcement and hence becomes visible only at times of crisis.

One such critical moment occurs when adult males change position. We have two reports of rank challenges between male bonobos at Wamba; in both cases, females were the deciding factor.[9]

Koguma Outranks Ude

The son of a powerful female, Aki, had begun to enter adulthood. This male, named Koguma, one day challenged the second-ranking male, Ude. Dragging a branch and screaming, Koguma charged straight at Ude, rushing narrowly past him. Ude leapt up and slapped Koguma in defense. The alpha male then intervened by mounting Ude, calming him down with rump-rump contact.

After a while, Koguma charged again. Because Ude counterattacked, the two males ended up flying about between shrubs and bushes, exchanging violent blows. When Koguma launched another attack, his mother, with an infant clasped to her belly, came to assist him. Ude fled as Aki, with the loud calls of other females behind her, chased him off. Koguma did not let up: he attacked Ude twelve times in the span of nine minutes. Every time Ude rose to retaliate, Aki would go after him. Towards the end, Ude became silent and avoided Koguma. Eventually, he fled to the nearest tree whenever Ude branch-dragged in his direction.

After that, Ude seemed unsettled. If Koguma charged, he would flee, or present his behind in an attempt to pacify the younger male.

Ten Outranks Ibo

The oldest and highest-ranking female, Kame, had three sons, the oldest of whom, named Ibo, was the alpha male. The beta female's son, named Ten, began to challenge Kame's sons, although he was usually defeated by Ibo. During this period, Kame was weakened by old age and did not intervene in her sons' affairs anymore. The mother of the challenger, in contrast, was in good health and began to attack Kame's sons. Once she even defeated Ibo himself in a serious physical battle.

The most critical confrontation took place, not between the males, but between their mothers. The two females had a hand-to-hand fight in which they rolled over the ground, and in which Kame was held down. Thereafter, such fights occurred repeatedly, but Kame never regained dominance.

Not only did the beta female rise to the alpha spot, her son did so, too. Ibo became submissive following his mother's defeat. The sons of Kame remained middle-ranking for a couple of years, but became quite peripheral after their mother's death.

Had Kame's sons been chimpanzees, they would no doubt have banded together to defend their positions collectively. In bonobos, however, male alliances are little developed, which allows females to exert much greater influence. As a result, a relatively young adult male can reach a top position provided his mother is of high rank. On the other hand, males whose mothers are over the hill, or dead, tend to drop in rank.

This brings us to perhaps the most puzzling aspect of bonobo society: females often dominate males. With a few notable exceptions, such as spotted hyenas and the lemurs of Madagascar, male dominance is the standard mammalian pattern. The reason is not hard to guess: males usually outweigh females and possess weapons, such as horns, tusks, or fangs, that are absent or much reduced in females. Because bonobos show the same kind of sexual dimorphism, albeit somewhat less pronounced than in many other primates, dominance by the "weaker" sex constitutes a huge violation of every biologist's expectations. The first time I noticed female dominance, I considered it an oddity. But the more captive colonies I studied, visited, or heard about, the more the pattern emerged as the norm rather than the exception.[10]

Could it be that cultural sensitivities surrounding the relation between men and women in our own societies led to a period of denial of this unique arrangement in one of our closest relatives? The situation remained unclarified until 1992, when the silence ended with a bang at an international meeting of primatologists in Strasbourg, France. Several students of captive bonobos reported observations and experiments that left little doubt about the issue. Amy Parish induced food competition in identical groups (one adult male; two adult females) of chimpanzees and bonobos at the Stuttgart Zoo. Honey was provided in a place from which the apes could collect it by dipping sticks into a small hole. The male chimpanzee would make a charging display through the enclosure and claim everything for himself; only when he was satisfied would he let the females fish for honey. In the bonobo group, on the other hand, the females would approach the honey mound together and en-

Because bonobos spend much time in the dense upper forest canopy, they are extremely hard to detect and follow.

gage in mutual sexual contact. After this, they would feed side by side, taking turns with virtually no competition. The male could make as many charging displays as he wanted; the females would be unimpressed.

Similarly, observers at the Belgian animal park of Planckendael reported that if a male tried to harass a female, all the females would band together to chase him off. That such behavior is not restricted to captivity is evident from observations at Wamba. According to Kano, males sometimes provoke counterattacks from a mass of females: "A group of males will not attack a female, but the opposite can occur."[11] At the center of a traveling party, one usually finds high-ranking females close together. Their sons are allowed to enter this aggregation, but adult males without mothers tend to stay at the periphery. The picture emerging from Wamba, then, is one of a female-centered society, in which even the male rank order is largely dictated by mothers.

At the meeting in Strasbourg, Furuichi further described what looks very much like female dominance in relation to food: "Males usually appeared at the feeding site first, but they surrendered preferred positions when the females appeared. It seemed that males appeared first not because they were dominant, but because they had to feed before the arrival of females. Even middle- and low-ranking females could displace males."[12]

Since females rarely resort to aggression, the best evidence for their high status is the way they control highly prized foods. Hohmann and Fruth documented food-sharing among bonobos at Lomako, which occasionally revolved around meat, but more commonly around large *Anonidium* and *Treculia* fruits. The possessor of the food was almost always an adult female. Bystanders surrounded her, some begging by stretching out a hand or touching her mouth. Males would display in the vicinity by breaking off branches and charging about, or they would hang around at the edges of feeding clusters. They would be nice to the young, all of whom had free access to the food. Not permitted to enter the cluster themselves, the males could do little aside from steal scraps from infants. Even if a male was the first possessor, he often lost the food to one of the older females. Perhaps as a result, males lacked the confidence to share themselves and held on tightly to whatever they had managed to get their hands on.

The image of female-controlled food distribution with waiting and "parasitizing" males at the margin is dramatically different from the typical feeding clusters of chimpanzees, in which an adult male holds an animal carcass while

(continued on page 82)

INTERVIEW WITH BARBARA FRUTH AND GOTTFRIED HOHMANN

In September 1995, I interviewed the ethologists Barbara Fruth and Gottfried Hohmann, who have worked in Lomako Forest since 1990. This German couple are currently affiliated with the Max-Planck Institut für Verhaltensphysiologie, in Seewiesen, Germany, and Miami University in Oxford, Ohio, in the United States.

FDW: What kind of research is currently going on at Lomako?

HOHMANN: Barbara's project on nest-building has been completed. We now wish to do field ecology, such as food competition between bonobos and other primates. Lomako can fill a special niche in this regard as our apes are unprovisioned; such a project would hardly make sense at Wamba. Further, there is the genetics project: comparing DNA extracted from feces. Hopefully, we can determine paternity and genetic relatedness to see, for example, if males are more related to one another than females. If bonobo males are indeed philopatric, one would predict such an outcome.

FDW: How hard is it to habituate apes without provisioning?

FRUTH: My impression from talking with Christophe Boesch, who habituated chimpanzees without provisioning, is that our bonobos were actually easy. Nevertheless, at the beginning we had a hard time. The first year, we could see them only if there was a large group in a fruiting tree. As soon as they came to the ground, we lost them. But during the *Pancovia* season, we had a breakthrough: they gathered in lower fruit trees and saw us up close. Males were generally braver than females. Presently, we can see everyone, and even some females tolerate focal observations [in which a fieldworker follows a single individual for a period of time]. But it took at least three years before we got to this point.

HOHMANN: Our policy is never to push the animals, to always stay close enough to see them, and then wait for them to approach. The result is that we can get as close as ten meters. If we wait, some of them approach us to within three meters. Sometimes juveniles try to provoke us, just out of curiosity. They move from branch to branch till they are right above us, then they throw a branch or urinate on our heads and see how we react. But generally, the bonobos seem less interested in us than we are in them.

FDW: I will ask you the same question as Drs. Kano and Kuroda. Are female bonobos dominant, equal, or subordinate to males?

FRUTH: Adult females are dominant in every possible way. Even younger females sometimes dominate adult males. I once saw a grown male try to steal an *Autranella* fruit from an adolescent female, but she chased him furiously.

HOHMANN: There probably is female cooperation involved. Just to take Barbara's example, when this young female chased the male, there were other females around. They all screamed and yelled. I don't think that on her own she would have had a chance against the male.

FDW: Do the Lomako bonobos hunt or fish?

HOHMANN: Although we have seen attempts, such as apes running after a duiker [a small forest antelope] which rushes away, we have never seen them actually hunt or grab an animal. However, we have heard the duiker screaming and were on the spot in the same minute. Some duikers they were seen sharing must have been about ten kilograms in size, which would mean they were adult.

FRUTH: There is no fishing, but perhaps they eat a sort of shrimp. They walk through the water and stare into their hand while they let the water flow away through their fingers. Then they eat something. There are tiny, transparent crustaceans in the water. The local people also eat them. It's a delicacy. The apes sometimes wade for hours through the stream bed.

FDW: Bipedally?

FRUTH: [*Laughing, knowing the wild idea I'm referring to.*] No, they simply wade through the water on all fours; these little rivers are very shallow.*

FDW: Barbara, bonobos are not the only apes to build nests. Is there anything special about them? [A few weeks earlier, Fruth had successfully defended her dissertation on nest-building.]

FRUTH: Bonobos often make nests by pulling small trees towards each other and weaving their branches together, resulting in a very springy platform. This way they can choose exactly where in three-dimensional space they wish to be. This is important, because the social organization is reflected in the nesting arrangement. Females are the first to build, usually high up in the trees. Others follow, with the adult males the last.

---

*I had earlier published some tongue-in-cheek speculations about how bonobos might fit the "Aquatic Ape" theory, according to which our ancestors began to walk upright when wading into shallow water (de Waal 1989, 182–86). Reports had suggested that bonobos enter streams bipedally. Fruth and Hohmann, however, have never seen such behavior.

They occupy lower levels and keep maximal distance to members of their own sex, while trying to get close to the females. Some of these features of nest-building may be bonobo-specific, others probably not. However, a striking difference exists in the aggregation at nest sites.

Whereas chimpanzees nest in clusters the size of their last traveling party, bonobos call each other in the late afternoon so that many of them come together at the nesting site. We have seen the entire community sleep at one site, in twenty-six nests. Could it be that they come together to exchange information about feeding sites? One party may know a large fruiting tree that they then go back to the next morning with the others in tow so that they can clean it out before the mangabey monkeys get there. This is one of the many unsolved questions.

FDW: The Wamba investigators report remarkably relaxed intercommunity relations. Do you have the same experience?

FRUTH: My first impression was of much greater violence. I once saw males of different groups wildly chase each other through the undergrowth with all females hanging in the trees, shouting and screaming. It looked so aggressive that I feared for my own life: I had goose bumps! I did not notice any injuries in the bonobos though. More recently, Gottfried has seen encounters more similar to what the Wamba team describes.

HOHMANN: It starts out very tense, with shouting and chasing, but then they settle down and there is female-female and male-female sex between members of the two communities. Grooming may occur, but remains tense and nervous. I have not seen friendly contact between adult males.

In other situations, the same groups may not get along. One day, I was following our study group when I was surprised by sudden drumming sounds right behind me. The bonobos dropped down and rushed towards the others, resulting in lots of vocalizing and drumming. Then at the border of their territory, I found them sitting in the tree making aggressive displays and shouting. That day, they did not tolerate the other group.

FDW: This is the first time I hear about drumming bonobos. What is it like?

FRUTH: [*Jumping up and drumming on the table.*] It is brief, fast, and loud, but not like the elaborate concerts of male chimpanzees. Bonobos often jump with both hands and feet against a tree buttress.

FdW: Are you worried about the future of the bonobos at Lomako?

Fruth: Yes, we are concerned. At the moment the pressure on wildlife in the Lomako region seems low. In a forest not too far away from our site, where bonobos were studied by others only five years ago, people are now digging for diamonds. Similar things can happen any time at any place: Lomako could be next. The human population is growing, and as a result the demand for bushmeat and other forest products is high. Although bonobos may not be the specific target of hunters, degradation of the habitat may have the same effect.

We are less concerned about the timber industry now. A large German veneer manufacturer with a 99-year lease abandoned the concession because it was not profitable. The wood they wanted was too hard to get, and they left the area to look for something better.

others beg for a portion. Chimpanzee possessors share both with their fellow hunters, usually other males, and with females. Their tolerance is remarkable, as well as understandable for two reasons. Sharing with other males serves to cement political ties and to reinforce hunting cooperation; why should other males assist in the strenuous capture of a monkey if not for a piece of meat at the end? Apart from such reciprocity among male hunters, offering food to females may work out as a paternal investment: females share food with their offspring, some of whom may be the hunter's progeny. So a male hunter sharing with females may be indirectly feeding his young. Finally, some of the sharing with females is probably repaid with sex, because male chimpanzees are particularly generous to sexually receptive females.

In comparison, why should bonobos share? So far as we know, there are no hunting teams, so incentives for cooperation are unneeded. And since females often control the food, there is also less reason for sharing between the sexes; when a female offers food to a male, unless it is her son, this does not translate into a parental investment. The only remaining reason, then, is the cementing of political ties. In the bonobo's case, this probably applies not to so much to males as to the senior females. Kuroda found indeed that the category least prepared to share were adult males among themselves.

Bonobo infants are born small and develop slowly compared to other apes. This infant is already more than two years old.

Possibly, then, the range of return benefits associated with sharing is more limited in the bonobo than in the chimpanzee. It may sound paradoxical that the more dominance-oriented and violence-prone species may actually have more reasons to share food, but it makes sense given the natural conditions

On their way back to camp after a long day in the forest, Kano (right), Takeshi Furuichi (next to Kano), and Chie Hashimoto chat with villagers.

under which chimpanzees engage in this behavior. Detailed data on food-sharing in wild bonobos and chimpanzees are needed to verify this hypothesis. The only existing comparisons concern captive apes. My data support the above view: I found chimpanzees to be more tolerant than bonobos in relation to food.[13]

This brings me to a final point: the tendency to romanticize bonobos. Even if strikingly pacific, they are not the long-lost noble savages. All animals are competitive by nature and cooperate only under specific circumstances and for specific reasons, not because of a desire to be nice to one another. The question of why bonobos are egalitarian and tolerant thus needs to be balanced with the question of in which areas they are most competitive. Such areas must exist: the history of biological and anthropological research warns strongly against idealization of a particular species or human culture. Science has erred before with a range of so-called peaceable species, from gorillas to dolphins, as well as with hunter-gatherer societies claimed to be free of aggression. Generally, such idealizations mean that something highly significant has been overlooked or, worse, covered up.

Thus, if wild bonobos frequently show physical abnormalities, such as deformed digits or even entirely missing hands or feet, we should seriously consider the possibility of trauma due to violence. There exists a distinct sex bias in the incidence of these abnormalities (adult males being the most afflicted category), and we know that male bonobos are involved in more fights than females. I speak from experience with naive claims: there was a time when I believed chimpanzees to be absolutely wonderful at managing conflict. Then it became known that males in the wild engage in brutal warfare against other groups, that they may kill and cannibalize infants of their own species, and that even within the group, violence may escalate to the point of deadly injuries. In the Arnhem Zoo, one male was fatally maimed and castrated by two others, and in Gombe National Park, a male chimpanzee would no doubt have lost his life had he not been treated by a veterinarian. This victim, too, suffered scrotal damage from an attack by group mates. I would not reject out of hand the possibility of similar aggression among bonobos.[14]

While we should think twice before attributing to bonobos violence that has not actually been witnessed, they are no saints. It is unlikely, given their spirited temperaments, that the harmony reigning in their societies is based entirely on inborn pacifism. Competitive undercurrents are not hard to detect. No matter which sex dominates, there must be negative sanctions to keep subordinates in line: injuries in zoo colonies point in this direction.[15] We have also seen indications from the field that females are serious rivals when it comes to their sons' dominance ranks, and that their fights can be vicious. In other words, bonobo society is not all rosy. The species is no exception to the rule that cooperative tendencies are best understood in conjunction with competitive ones, even though I agree that in bonobos the emphasis seems to have shifted to the former.

BONOBOS IN THE MIST

Bonding among female bonobos is surprising inasmuch as it violates a general rule outlined by the British anthropologist Richard Wrangham. Strong bonds between same-sexed partners normally concern the sex that stays all its life in the natal group. For example, male bonding of chimpanzees follows naturally from males remaining in the community into which they were born. The same is true of female bonding in Old World monkeys, such as macaques and baboons; males being the migratory sex in these species, females stay together, forming complex kinship networks. Bonobos are unique in that the migratory sex bonds with same-sex strangers later in life. Establishing an artificial sisterhood, female bonobos can be said to be *secondarily* bonded (if we consider kinship ties the primary bonds).

Not that everyone agrees that female bonobos deserve to be called "bonded." There are basically two schools of thought. Wrangham himself has depicted bonobo females as tolerant yet unbonded. Frequent grooming and sexual contact among females does not necessarily reflect mutual attraction; it may serve

BONOBO PARTY COMPOSITION AND SOCIAL ORGANIZATION

The following outline is based on field studies at Wamba, Lomako, and other sites in Zaire.

1. Like chimpanzees, bonobos live in male-philopatric fission-fusion societies.
2. Bonobos emphasize female bonding but show a potential for male bonding, whereas chimpanzees emphasize male bonding with a potential for female bonding.
3. Male kinship ties focus on the mother rather than on brothers.
4. Food occurs in large enough concentrations to allow multiple bonobos to forage together. Bonobo parties are typically mixed.
5. Bonobos are gregarious: one sees large parties at Wamba and nightly fusions in Lomako.
6. The female hierarchy, based on age and residency, is rather vague.
7. Males compete fiercely over rank, which is influenced by their mothers.
8. Females can monopolize prized foods; they often dominate males.
9. Despite hostility between groups, there is also peaceful mingling.

to reduce tensions rather than to cement ties. This assessment of bonobo females as associated yet unaffiliated leaves existing theories about the evolution of social bonding intact. The second school, most outspokenly represented by Amy Parish, argues that bonobo females are bonded by any standard; no other term will do. They not only tolerate but actually prefer one another, whereas males (except sons) are only marginal players in their lives. This school considers relationships among females as the most fundamental bonds in bonobo society, even if they are formed at a later age and do not overlap with genetic relations.

Although we know now *how* these close relationships are established—through the use of sexual contact and grooming—science has no answer yet to the question of *why* bonobos and chimpanzees differ in this regard. On the assumption that female grouping patterns are the key, the answer has been sought in different ecological conditions that permit females to forage together without too much competition.

Richard Malenky and Frances White, in collaboration with Wrangham, have compared the feeding ecology of the Lomako bonobos with that of chimpanzees. They conclude that bonobos have access to larger fruiting trees, which allow more individuals to feed together, and that they consume larger quantities of THV, a common food source in their range. They bite open the tough fibrous sheaths and chew the soft pith of canelike herbs. The investigators believe that the social structure of bonobos reflects the presence of these predictable and abundant food sources, on which multiple females can feed without conflict. In chimpanzees, in contrast, females are forced to seek their food independently because it is more dispersed.

Since herbaceous foods are a staple of another close relative, the gorilla, Wrangham has speculated that the total absence of gorillas in the bonobo range may have opened an ecological niche that is simply unavailable to the chimpanzee (chimpanzees and gorillas are sympatric in large parts of their ranges):

> In the Middle Pliocene, two or three million years ago, chimpanzee- and gorilla-like apes lived together on the left bank of the Zaire River. A brief cold, dry period around that time led to the loss of the perennial herbs and therefore also of the gorillalike apes, which depended on the herbs for food. When the humidity increased, the chimpanzeelike apes continued to eat fruit but also found themselves released from feeding competition with the gorillalike apes. Whenever the fruit trees failed to provide abundant food, these protobonobos could turn to the abundant herbaceous gorilla foods on the forest floor. Rather than being forced by feeding competition into small parties when tree fruits were scarce, the protobonobos could continue to forage together.[16]

However plausible this scenario may sound, it is not without its problems. For one thing, it assumes that bonobos descended from chimpanzeelike ancestors,[17] whereas, as we have seen, other scientists believe bonobos to represent the ancestral form. It also raises the question of why, if ecology dictates social organization, bonobos did not adopt a more gorillalike social organization in which a number of females are led by a male who competes with other males for their "possession." The bonobo seems to have followed almost the opposite evolutionary path: large domineering males are nowhere in sight. Perhaps the chances for a gorillalike system were slim from the beginning, since abundant foods in large "patches" allowed females to associate, bond, and eventually form alliances that held male ambitions in check. If so, a slight initial advantage in the battle of the sexes may have allowed bonobo females to achieve a totally different social organization, in which not males but they themselves were in charge.

The evolutionary background of the unique bonobo society remains shrouded in mist. Further ecological studies will no doubt help answer some of the questions, but we are not even near a reconstruction of the divergence among the African hominoids, including ourselves, and the environmental pressures that caused it. Some changes may follow logically from others, but so long as we do not understand the key transitions, we are far removed from plausible scenarios. One highly problematic issue, not even mentioned thus far, is why there is so little competition between groups of bonobos. Chimpanzee males are known to kill each other over territory, gorilla males occasionally fight to the death over females, and our own species has a long history of battlefields scattered with the bodies of thousands of men. Bonobos, in contrast, seem merely to "visit" their neighbors, with some hostility and tension, but no murderous intent.

The first peaceful intergroup mingling was observed in 1979 at Wamba, where two different communities came together and stayed together for a week. At a recent meeting, Kano played a video of such mingling. First one sees bonobos fiercely chasing each other, screaming and barking, but without physical fighting. Then, gradually, females of the different groups engage in sexual contact and even groom one another. In the meantime, their offspring play with those of the other group. Even a few males of different groups approach one another to engage in a brief scrotal rub. Those familiar with the brutal encounters between chimpanzee communities, described in gruesome detail by Jane Goodall, can only shake their heads in wonderment at bonobo intercommunity relations.

Idani, who recorded 32 separate intergroup encounters at Wamba, characterizes the typical interaction between males and females of different groups as sexual and friendly, whereas males are hostile and standoffish towards males of another group. Copulations between males and females of different groups are common during the first fifteen minutes of an encounter. Provisioning may be partly responsible for these group mergers, since many occurred at the feeding site.

Before ascribing the bonobo's inoffensive intercommunity relations to human influence, however, we should consider that for chimpanzees, exactly the opposite has been argued. Some anthropologists dismissed the brutal warfare in Gombe as an artifact of provisioning, because, they claimed, concentrated food sources give rise to violence. If true, then why should bonobos under similar circumstances fight so little? Idani adds that encounters away from the feeding site support his conclusions: the same temporary fusions occurred in the forest. Furthermore, the latest reports from Lomako, where provisioning is absent, suggest the same kind of relaxed intergroup relations (see the interview with Barbara Fruth and Gottfried Hohmann on pp. 79–82).

The extensively overlapping travel ranges of bonobo communities and direct observations of relatively peaceful mixing suggest that bonobo intercommunity relations are strikingly different from those of their closest relatives. When the mist that shrouds the evolutionary pressures that shaped bonobo society lifts, we may perhaps get an answer to the question of how they managed to escape what many people consider the worst scourge of humanity: our xenophobia and tendency to exterminate enemies on a large scale. Could it be that it is because bonobos do not fight for a fatherland but, if they fight at all, for a motherland?[18]

LIFE IN THE FOREST

Photographs of bonobos in their native habitat are exceedingly rare, and fewer still capture the peculiarities of their social life. Here a high-ranking female grooms her young adult son. He will follow her through the forest and perhaps one day rely on her support to usurp high status among the males. OVERLEAF: *These bonobos gaze intently in the direction of calls that come from the right. The apes' movements are carefully coordinated with what they know about the location of other group members and of other groups.*

After a period of clinging to their mother's belly, older infants ride jockey-style on her back. They nurse for four years and are carried around even longer; their mother takes them everywhere she goes. In the fluid society of bonobos, females with young often encounter others like themselves.

With its mother feeding beside it, an infant sits in a giant buttressed tree (above). A juvenile male (opposite) utters the characteristic high-pitched long-distance calls that bonobos use to announce their presence to others.

CHAPTER 4

APES FROM VENUS

During a recent visit to the Wild Animal Park, northeast of San Diego, I provided old acquaintances with a sharable meal, while a camera team recorded their table manners for a popular science program. The bonobos did exactly what they were supposed to do: they resolved tensions over food with sex.

We filmed them in their spacious, grassy enclosure with palm trees, in the dry, warm weather of southern California, which suits these apes surprisingly well, given that their natural habitat is moist and humid. Loretta recognized me straightaway and, in her usual fashion, turned her behind towards me, staring at me from between her legs with bright eyes, while stretching out a foot in invitation. Even though there was a full-grown, muscular male in her group, Loretta was the uncontested queen. At the time, the rank order was as follows: (1) Loretta, a 21-year-old female; (2) Akili, a 15-year-old male; (3) Lenore, a 13-year-old female; and (4) Marilyn, an 8-year-old female.

When a bundle of ginger leaves, a favorite food, was thrown in front of them, Loretta immediately seized it. After a while, she allowed Akili to eat from her leaves, but Lenore remained reluctant to join. This was not because

Sex is the glue of bonobo society. This female squeals while being mounted by a male. Not all contacts are between members of the opposite sex, however; sex occurs in virtually all partner combinations and in a unique variety of positions.

of Loretta, but because Lenore and Akili for some reason do not get along. The caretaker told me that this had become a long-standing problem in the group. Lenore kept looking at Akili, avoiding his every move. These troubles were eventually resolved sexually. Lenore presented from a distance a number of times. When Akili failed to respond, she approached him and pressed her genital swelling against his shoulder, rubbing slightly. After this, she was allowed to join the group without further problems, and all of them peacefully ate together from the food that Loretta held firmly in her hands.

The adolescent of the group, Marilyn, had something else on her mind. She was enamored with Akili, following him around, sexually inviting him many times. Marilyn played quite a while in the pool, manually stimulating her genitals while dipping her lips in the water. After having excited herself in this manner, she pulled at Akili's arm and led him by the hand to the water for a copulation. Akili obliged several times, but was clearly torn between Marilyn and the food bonanza. I wondered if Marilyn had developed a water fetish. Or was this just a temporary variation on an old theme? Bonobos seem to put a lot of imagination into their sexual adventures.

In the meantime, Loretta showed great interest in Lenore's baby. Whenever the infant came close, she would briefly stimulate its genitals with a finger, followed by a belly-to-belly embrace during which she thrust against the infant. One time, the mother stimulated Loretta's genitals, after which she pushed her offspring towards her, as if she wanted Loretta to hold it.

In this short stretch of time, we thus saw bonobos use sex for sex (Akili and Marilyn), for appeasement (Lenore and Akili), and as a sign of affection (Loretta with the infant). Perhaps we should not call all of this "sex," because people tend to think of it as a self-contained behavioral category aimed at an orgasmic climax. We associate sex with reproduction and sexual desire, whereas in the bonobo, it mixes with all sorts of other tendencies. Gratification is by no means always the objective, and reproduction is only one of its many functions. On the other hand, human sexuality, too, may have a broader significance than commonly acknowledged. Our sexual urges are subject to such powerful moral constraints that it may have become hard to recognize how—as Sigmund Freud was the first to point out—they permeate all aspects of social life, and not only with regards to partners to whom we admit to being sexually attracted. Some of these moral constraints exist for good reason, and I am not questioning them, but bonobo society may quite possibly provide insight into how our sexuality might function without them.

Face-to-face mating used to be seen as proof of the dignity and sensibility separating civilized humans from so-called subhumans. The frontal copulatory position was elevated to a cultural innovation of great significance, one that fundamentally altered the relationship between men and women. It was felt that preliterate people would greatly benefit from education about this mode of intercourse, hence the term *missionary position.* In the 1960s, American cultural anthropologists debated its advantages:

> Our guess is that it changed, for the adult female, the relative roles of the adult male and of the infant, since after the innovation there is a much closer similarity for her between her reception of an infant and of a lover. This may have helped to spread the tender emotions of mammalian mother-infant relations to other interpersonal relationships within the band, ultimately with such further consequences as the Oedipus complex.[1]

> It seems not only that the adult male becomes in face-to-face copulation, a surrogate suckling to the adult female by virtue of his position; but also that the adult female becomes a surrogate suckling to the adult male by virtue of her behavior, which is that of soliciting and receiving a life-giving liquid from an adult bodily protuberance.[2]

While I find it hard to read such armchair (or bedroom) theories with a straight face, they represent serious attempts by social scientists to set human sexuality apart from that of animals. To every biologist, however, it is crystal-clear that sex is one area of human behavior in which cultural experimentation is about as constrained as it is with breathing. Where would we be without the occasional meeting between sperm and egg? There are only so many ways to arrange these meetings: the hormones urging us to unite with the opposite sex, and the anatomical features making these acrobatics both feasible and pleasurable, are biologically dictated. It is also obvious from our frontally oriented genitals that the missionary position, albeit nonobligatory, has been favored by natural selection. Antedated by billions of years of sexual reproduction, the effects of civilization on the sex act are marginal at best.

Few sexual patterns typical of our species are absent in bonobos. As Claudia Jordan put it: "There is hardly any practically possible mating position that would not occur."[3] The best way to convey the richness of this ape's sexuality is simply to list the patterns observed at the San Diego Zoo. Before I went there, I had heard that bonobos were sexy, but I was nonetheless amazed by the sheer variety of positions and the extent to which the apes mutually stim-

Face-to-face copulation was long thought to be uniquely human. Bonobos, however, are acrobats who practice every possible mating position. Their genitals are adapted for frontal intercourse, which is common among them. (Photo by Frans de Waal.)

ulated each other. The most common mating pattern is the ventro-dorsal (belly-to-back) position. This is also the typical position of most primates: chimpanzees, for example, almost exclusively mate "doggy style." Bonobos share with us, however, that their genital anatomy seems actually adapted for the ventro-ventral (belly-to-belly) position.

The female's genital tissue, mainly the labia and clitoris, periodically swells up to balloon size, indicating sexual receptivity. Swelling and vulva are located more between the legs than in the chimpanzee, and the clitoris is prominent, erectile, and likewise frontally oriented. Given this anatomy, it is not surprising that females seem to prefer the frontal position, which guarantees optimal stimulation. Male evolution may have lagged behind, resulting in a mismatch between male and female preferences. Females invariably invite males lying on their backs, and sometimes change towards this position if the male has started out differently, yet ventro-dorsal matings occur about twice as often as ventro-ventral matings. The most important point, though, is that thus far all investigators have reported the regular use of both positions, which means that both can be said to be species-typical.[4]

Also characteristic is pseudo-copulation between females mounted ventro-ventrally, one female carrying the other. This posture—in which one female may be lifted off the ground by the other while she clings to her, much as an infant clings to its mother—allows both females to make sideway movements. The females rub their clitorises together with an average of 2.2 lateral moves per second; the same rhythm as that of a thrusting male. The pattern is widely known as *GG-rubbing:* an abbreviation of genito-genital rubbing first proposed by Kuroda. It has been observed by all students of bonobo behavior and is unique to the species.

Females also sometimes mount each other with both partners facing in opposite directions. While one female lies on her back, the other stands over her, with her back turned, rubbing her genitals against her partner's. A comparable but less intense back-to-back posture occurs between males, with both partners standing on all fours briefly rubbing their rumps and scrota together. This is known as *rump-rump contact.*

In contrast, the posture during so-called mutual penis rubbing resembles that of a heterosexual mount, with one male (usually the younger) passively on his back, the other male thrusting on him. Because both males have erections, their penises rub together. I have never seen ejaculations during sex between males nor attempts at anal penetration. Kano further describes *penis fencing,* a rare behavior, thus far seen only at Wamba, in which two males hang face to face from a branch while rubbing their penises together as if crossing swords.

During my original studies at the zoo, the same Lenore who is now starting her own family at the Wild Animal Park was still an infant. Like all young bonobos, she was fascinated by sex, jumping on top of adults engaged in it and sometimes pressing her vulva against her mother's swelling when she GG-rubbed with other females. This way, Lenore partook in whatever went on and learned the various contexts in which sex plays a role. She also initiated sexual games of her own, mainly with willing adolescent males. She would climb on a male's belly pressing her body against his genitals, whereupon the male—either in a sitting or reclining position—would make a series of thrusts. Mounts with this infant never resulted in intromission or ejaculation.

Apart from sexual behavior, I saw patterns that are perhaps better classified as erotic in that, even if adults of the opposite sex were to engage in it, reproduction could not possibly result. The first is mouth-to-mouth kissing in which one partner places his or her open mouth over that of the other, often with extensive tongue-tongue contact. While typical of the bonobo, such "French-kissing" is totally absent in the chimpanzee, which engages in rather platonic kisses. This explains why a new zookeeper familiar with chimpanzees once accepted a kiss from a male bonobo. Was he taken aback when he suddenly felt the ape's tongue in his mouth!

Another erotic pattern is fellatio, that is, one partner taking the penis of another in the mouth and stimulating it. This happened regularly during rough-and-tumble play among juveniles. Chasing and wrestling would be interrupted by erotic games in which all juveniles might participate, some of them mounting, others engaging in oral sex. Play would resume within a few minutes.

The final erotic pattern is manual massage of another's genitals. In the majority of instances, this was done by the adult male to one of the adolescent males. The younger male, with back straight and legs apart, would present his erect penis to the adult male, who would loosely close his hand around the shaft, making caressing up-and-down movements. This is the social equivalent of masturbation, which bonobos also engage in. In the males, neither genital massage nor masturbation was ever observed to produce ejaculation. The most regular masturbators were adolescent males and adult females.

The latter is significant because of yet another uniqueness claim: from Frank Beach to Desmond Morris, scientists have declared female orgasm to be exclusively human. While people readily assume males to enjoy sex, many appear skeptical about females. This reflects the puritan belief, prevalent until early this century, that sex is a man's privilege and a woman's chore. To assume that female sexual arousal is limited to our species is to deny it the same biological roots as male arousal. If female bonobos habitually masturbate, however, this activity must surely produce enjoyable sensations. Otherwise, why would they do it? We also know from laboratory experiments with stumptail macaques—another primate with a highly developed sexual repertoire—that we are not the only species in which females at the climax of copulation experience an increased heart rate and rapid uterine contractions. These monkeys, and probably many other primates as well, fit the physiological criteria of orgasm as defined by Masters and Johnson.[5]

If the sounds and facial expressions of bonobos are any indication, not only masturbation, but also sexual intercourse must be quite gratifying. Females frequently bare their teeth in a pleasure grin during coitus, particularly towards the end when the male slows down for his final, deeper thrusts. Furthermore, females often utter characteristic screams and squeals before or during coitus, as well as when they engage in GG-rubbing with other females. Sexual partners often face each other, so they can closely monitor each other's facial expressions and sounds, and the exchange becomes quite intense and intimate. Indeed, sexual activities may be interrupted in case of a mismatch of emotions, as indicated by an early study at the Yerkes Primate Center by Sue Savage-Rumbaugh and Beverly Wilkerson:

> Slow motion study of videotaped copulatory bouts indicated that, in many cases, the speed and intensity of thrusting was visibly altered or terminated as a function of changes in the facial expression or vocalizations of one of the participants. These observations strongly suggest that the pygmy chimpanzee is responsive not only to his own physiological feedback during copulation, but also to the subjective experiences of the partner as mediated via facial expressions and vocalizations. On numerous occasions, either the male or female was observed to terminate thrusting when the partner could not be engaged in eye contact or otherwise indicated disinterest by yawning, self-grooming, etc.[6]

Lest this overview of the sexual and erotic behavior of bonobos leave the impression of a pathologically oversexed species, I must add, based on hundreds of hours of watching bonobos, that their sexual activity is strikingly casual and relaxed. It seems a completely natural part of their social life. Also, even though the bonobo is a serious contender for the title of sex champion of the primate world, its sexiness should not be exaggerated. Bonobos do not, in fact, engage in sex all the time. At the zoo, the average bonobo initiates sex once every one and a half hours, whereas the average chimpanzee does so once every seven hours. In the wild, the frequencies are no doubt lower. Many of the contacts, particularly those with the very young, are not carried through to the point of sexual climax. The partners merely pet and fondle each other. Even the average copulation between adults is quick by human standards: 13 seconds at the San Diego Zoo, and 15 seconds at Wamba. Instead of an endless orgy, we see a social life peppered by brief moments of sexual activity.

ATTRACTIVE AT A PRICE

While the influence of economic factors and social pressures should not be underestimated, the bedrock of the human nuclear family is the bond between husband and wife. This applies regardless of whether a man has one or, as allowed in the majority of cultures, more than one wife.[7] This bond is sustained in part through regular sexual intercourse. Sex could never have come to play this pivotal role without a dramatic prolongation of female receptivity. If women, like most female mammals, engaged in sex only a few months each year, or only a few days each month, they might have a hard time winning male commitment. Of course, in modern societies, many women raise children without direct male involvement, but it is safe to assume that this option hardly existed for our ancestors. Surrounded by predators and enemies, and living a marginal existence, in which every source of subsistence counted, male sup-

Two adult males engage in rump-rump contact: they mutually rub their scrota together, in the male equivalent of the typical genital rubbing among females. In bonobo society, males are clearly the more competitive sex: the vast majority of aggressive chases at Wamba's feeding site are by males against males. Brief rump-rump contact serves as a conciliatory gesture. (Photo by Takayoshi Kano.)

port probably made a huge difference. It made it possible for protohominid females to raise more offspring than the apes from whom they had descended. It might be the chief reason why we, and not they, populated the world.

Although male apes cavort with youngsters, tolerate their antics, and sometimes protect them fiercely, it is fair to say that care of the young in our closest relatives rests squarely on female shoulders. After a gestation of around eight months, the mother nurses for four years and carries and protects her offspring for even longer. The maternal investment of apes is surpassed in magnitude only by that of a handful of other long-lived mammals, such as elephants, whales, and ourselves. Female chimpanzees give birth about once every six years, whereas bonobos at Wamba do so about every four and a half years. Even if this birthrate is slow compared to those of most other animals, it may be the maximum that apes in the wild can manage. Bonobo females at Wamba sometimes give birth so quickly after the previous time that they end up nursing two offspring. One sees the mother walking bipedally with an infant clinging to her belly and a juvenile riding on her back. This must be a heavy load; the species has probably stretched the single-parent system to the limit.[8]

Despite the absence of stable mate bonds, bonobos share with us a dramatically extended sexual receptivity. Females are most willing to engage in sex when they are maximally swollen; during this phase mating males also thrust faster—perhaps reflecting greater arousal. Increased receptivity has been achieved by extending the period of genital swelling. Whereas the chimpanzee has a menstrual cycle of approximately thirty-five days, the bonobo's is closer to forty-five days, and the period of swelling covers a greater portion of the cycle (75% compared to 50% in the chimpanzee). In addition, bonobo females resume swellings within a year after having given birth—when they are definitely not yet fertile—which further adds to the amount of time when they are sexually attractive to males. These characteristics make for quite a contrast: the chimpanzee female is receptive less than 5 percent of her adult life, whereas the bonobo female is so nearly half the time.[9]

As can be gathered from remarks by zoo visitors, most people are disgusted by the eye-catching genitals of apes. Some mistake them for abscesses, and the most confused reaction I once heard was from a woman who exclaimed, "Oh, my gosh, is that a head that I see?" Ape males, in contrast, know exactly what they see: for them there is nothing more exciting than a female with a voluminous pink behind. Personally, I am so used to these anatomical features that I do not regard them as weird or ugly, although "cumbersome" does come to mind. Fully swollen female bonobos cannot sit down normally; they awkwardly place their weight on one hip or the other. The swelling tissue is fragile, bleeding on the slightest occasion (but also healing rapidly). It is a heavy price to pay for being attractive.

Unfortunately, too, once swellings are established as an arousing signal in a species, they may never be lost. Imagine an ancestral species subject to periodic genital swellings evolving towards increased female sexual receptivity without a continuous swelling. We immediately see the problem: this would have required a transitional period in which females with reduced swellings would have been in competition for male attention with females with larger swellings. Given the long history of attractiveness based on swelling size, the first class of females would most likely have lost this battle even if their swellings had lasted a bit longer. As with the peacock's tail and the giant antlers of some prehistoric deer species, scaling back is not an evolutionary option if it compromises sex appeal.[10]

Since our species did achieve extended receptivity without swellings, it is thus safe to assume that our ancestors had none to begin with. We have avoided the toll that bonobos pay for evolution in the same direction. Women have reason to be glad about this. Had we gone down the same evolutionary path, allowance would have had to be made in the design of chairs, for example, to accommodate these large appendages.

The American anthropologist Owen Lovejoy argues that a strong sex drive and the ability to mate throughout most of the cycle enabled female protohominids to "purchase" male services. Popularizing this idea in her book *The Sex Contract,* Helen Fisher explains:

> But how to enlist male service? Some ancient females were sexier than others. They copulated throughout *more* of their monthly cycle, throughout *more* of their pregnancy, and *sooner* after delivering their young. These females, though burdened down with helpless young, attracted constant and close attention while they were in heat. During daytime expeditions they were in the center of the group. At night when everyone assembled to beg for meat, these females got the most. Thus sexy females were healthier and safer; their children were healthier and safer too. Therefore the children of sexy females disproportionally lived to adulthood and passed on the genetic propensity to copulate *throughout* the month, *during* pregnancy, and *shortly after* parturition. Protohominid females lost their period of heat.[11]

At this point it is unclear whether the bartering of sex for food that is central to these speculations suffices to explain the origin of the nuclear family. First, since other monogamous primates fail to show high sexual activity, frequent sex may not be necessary to hold a pair together. Second, it is hard to imagine that increased female receptivity could have lured men into relationships that did not carry advantages beyond sexual satisfaction. There must be reproductive benefits associated with these relationships, not just for the females, but also for the males. The successful raising of offspring is important to both sexes, hence rather than a female ploy, the male-female bond is best looked at as a mutual contract. As noted by Meredith Small in *Female Choices:* "Those who argue that females need to snare males into paternal care by giving them sex overlook the fact that the human males' reproductive success is as dependent on parental care as is female success. If a male abandons his offspring, it's likely to die. Thus females need not coax males to stay and help, selection will develop this behavior anyway."[12]

MAKE LOVE, NOT WAR

My original reason for studying bonobos had little to do with sex—or so I thought. I have a long-standing interest in aggressive behavior, and particularly in the way conflicts are resolved. For example, after a fight between two chimpanzees, they often come together to hug and kiss. Assuming that these reunions serve peace and social cohesion, I labeled them reconciliations. Any species that combines cooperation with a potential for conflict needs conciliatory mechanisms. Imagine how short marriages would be if people had no way to express regret at their actions or to make up for hurting each other. I found each primate species I studied to possess its own, unique manner of social repair (these findings are summarized in my book *Peacemaking among Primates*). In bonobos, sex turned out to be the key ingredient.

As soon as caretakers at a zoo approach bonobos with food, the males develop erections. Even before the food is thrown into the enclosure, the apes are inviting one another to sex: males invite females, females invite males, and GG-rubbing among females is also common. What is all this sex about? The simplest explanation would seem that excitement over food sparks over into sexual arousal, as if enthusiasm for food and sex get mixed up. This could well be the case, yet a third factor is probably the real underlying cause: competition.

In all animals, attractive food strains the relationships. There are two reasons to believe that sexual activity is the bonobo's answer to this circumstance. First, anything, not just food, that arouses the interest of more than one bonobo may trigger sexual contact. For example, if two bonobos approach a cardboard box given to them, they will briefly mount each other before playing with the box. I have even seen adult females GG-rub when one had only found a little piece of frayed rope and another hurried over for a closer look. Such situations often cause squabbles in other species, whereas bonobos are quite tolerant. They use sex to divert attention and change the tone of the encounter.

Second, sex often occurs in aggressive situations that have nothing to do with food. For example, after one male has chased another away from a female, the two may engage in a scrotal rub. Or when one female has hit a juvenile, and the juvenile's mother has come to its defense, the problem may be resolved by intense GG-rubbing between the two adults. Based on hundreds of such incidents, my study produced the first solid evidence for sexual behavior as a mechanism to overcome social tensions. This function is not absent in other animals (or humans, for that matter), but the art of sexual reconciliation may well have reached its evolutionary peak in the bonobo.

With this in mind, many encounters take on special significance. Once, in a group of juveniles, the oldest female, Leslie, found a younger male, Kako Jr. (Kakowet's last offspring), blocking her way on a branch. First she pushed him. Kako, who was not very confident in trees, did not move: he tightened his grip, grinning nervously. Next Leslie gnawed on one of his hands, presumably to loosen it from the branch. Kako uttered a sharp peep—but stayed put. Then Leslie rubbed her vulva against his shoulder. This calmed Kako, and he walked out in front of her. It seemed that Leslie had been very close to using force, but instead had reassured both the young male and herself by means of genital rubbing.

The dominant male of the colony, Vernon, regularly chased a younger male, Kalind, into the dry moat. It was as if Vernon wanted to keep Kalind out of the group. The young male always returned, only to be chased back by Vernon. After a number of such incidents—sometimes more than a dozen in a row—Vernon usually gave in. He would fondle Kalind's genitals, rub scrota with him, or engage in a rough tickling game. Without such friendly contact, Kalind would not be allowed to return. So, after emerging from the moat, his first task was to hang around the boss and await a cordial signal.

Bonobos thus substitute sexual activities for rivalries. Sex keeps competition down at feeding time and facilitates rapprochement in the aftermath of fights. Given these conciliatory and appeasing functions, it is not surprising that sex occurs in so many different partner combinations; the need for peaceful coexistence is obviously not limited to heterosexual pairs. In my study, sex occurred as often in combinations incapable of reproduction (such as between two males or two females) as in potentially fertile partner combinations. Furthermore, the latter category are only *potentially* fertile. Since adult females are fertile only a few days per cycle, not every copulation has the capacity of resulting in conception. Swellings are unreliable indicators of fertility; the swelling phase far exceeds the period of ovulation, and swellings also occur in pregnant and lactating females who are not ovulating at all. I estimated, therefore, that three-quarters of the sexual encounters in the colony had nothing to do with reproduction.

That food induces sexual activity instead of overt competition has been observed, not only in zoos and among the provisioned bonobos of Wamba, but also in Lomako. There, Nancy Thompson-Handler saw bonobos engage in sex when they entered trees loaded with ripe figs, or when one of them had caught a forest duiker. The flurry of sexual contacts would last for five to ten minutes, after which the apes would settle down to share and consume the food.

Occasionally, the role of sex in relation to food is taken a step further, a step that brings bonobos very close indeed to the scenarios of human evolution developed by Lovejoy, Fisher, and others. Females who—usually because of youth—are unable to dominate a male may employ sex as a "weapon." In my

Why else would this bonobo female be masturbating if not for pleasure? Bonobo females have unusually prominent clitorises and are among the most sexually solicitous creatures in the animal kingdom.

early study, when Loretta had not yet achieved the queen status of today, her self-confidence fluctuated with the size of her swelling. When swollen, she would not hesitate to approach the adult male, Vernon, when he had food. She would mate with him, uttering high-pitched sounds, while taking his entire bundle of branches and leaves. She hardly gave him a chance to pull out a branch for himself, sometimes grabbing the food from his hands in the midst of intercourse. This was quite in contrast to periods when Loretta had no swelling; then she would patiently wait till Vernon was ready to share.

I once shot a photograph of a young female grinning and squealing during copulation with a male who held two oranges, one in each hand. The female had presented herself to him as soon as she noticed what he had. She walked away from the scene with one of the two fruits. A response by a member of a professional audience to which I had shown this picture lampooned human familiarity with this pattern. Right after my presentation, when people were filing into a restaurant for lunch, a burly ethologist jumped onto a table holding two oranges in the air. He got lots of laughs; our species has a quick grasp of the sexual marketplace.

In his book *The Unknown Ape,* Suehisa Kuroda described the following events at the Wamba feeding site:

> A young female, named Mayu, came over to stare at the group's beta male, Yasu, who had several pieces of sugar cane. She turned around to present her swelling, slowly pushing it towards the male. Yasu, who had no erection, turned a little aside, but her behind followed him. Soon, he copulated with her. Looking at Yasu, she took one piece of sugar cane, and left. It was almost as if Mayu had forced Yasu to buy a copulation!
>
> To me it is always puzzling what males actually gain from these sexual encounters: in most cases the encounters are rather brief and do not seem to end in ejaculation. Nevertheless, it is certain that females "know" that sex produces enough tolerance in males to allow the females to remove food from their hands. They seem to seek sex for this purpose.
>
> Even three-year-old infants have learned this tactic. Kagi's infant once begged and obtained a bit of pineapple from a young adult male, Jess. She begged again, but Jess moved the pineapple out of reach. Then she turned to present her bottom, upon which Jess thrusted his erect penis a few times against her. The infant took a large piece of pineapple, and left him alone.[13]

The use of sex to promote sharing, to negotiate favors, to smooth ruffled feathers, and to make up after fights is enough to make it the magic key to bonobo society. On top of this, sexual attraction may explain the species' unique party organization. Remember how bonobo society is characterized by mixed traveling parties and female bonding. Integration of young females into the residential community is accomplished by frequent GG-rubbing and grooming. Sex reduces competition among females and allows them to travel and forage together. Females are much of the time interested in sex with each other as well as attractive to the males. There is almost always a receptive female present in these parties, which in turn guarantees male company. Chimpanzee males, too, consort with swollen females, but the females of their species are much less often in this state. In bonobos, the same attraction combined with prolonged sexual receptivity has resulted in almost continual male-female association.

It is impossible to know what came first—attraction between the sexes or attraction among females. If captive data are any indication, female bonding seems a basic feature of bonobo society. Observations by Amy Parish at San Diego's Wild Animal Park demonstrate a distinct preference of bonobo females for one another's company. Eight different groupings were tried out at the park, most of which included a single adult male, two adult females, and a couple of immatures. The adult partners varied: three different males and five different females figured in the rotations, but females always had a partner choice between adults of both sexes. Hundreds of records told us how much time individuals spent together, who approached whom, which partners groomed together, and so on. The conclusion was that females favored the company of members of their own sex. Females sat together, groomed each other, and played together considerably more than with the male in their group. They actively pursued these contacts: females followed each other around *seven times* more often than they followed the male. Because females also associated more with immatures (many of whom were their offspring), adult males tended to be rather peripheral to group life.

INTERVIEW WITH AMY PARISH

Amy Parish, who recently received her Ph.D. in anthropology from the University of California at Davis, studies female bonding in captive bonobos. She conducted her thesis research under the supervision of Sarah Blaffer Hrdy and myself. We discussed her findings and thoughts most recently in October 1995.

FdW: Do bonobos throw any new light on human social evolution?

PARISH: The bonobo social system offers an alternative to what is otherwise a rather bleak picture of opportunities for female bonding. When I began my studies, I expected to see strong bonds between males and females, because this is what fieldworkers reported at the time. I discovered that the bonds among females are actually stronger than those between the sexes. Moreover, females control access to preferred foods, sharing it with each other more than with males. Females form alliances in which they cooperatively attack and injure males: they dominate them. Inasmuch as bonobo females are generally unrelated, their society shows that kinship is not a requirement for female bonding.

Some people believe that women are an exception—that only women are capable of forming enduring friendships even if unrelated. Others believe that women aren't very good at bonding with one another at all. The bonobo is changing all this by showing that women are not the only bonded hominoid females.

FdW: You were one of the first to make a connection between the GG-rubbing, bonding, and power of female bonobos. How are these elements related?

PARISH: Evidence for female bonding and power is strong and growing. There is even evidence in the older literature; it is just that no one had put the whole idea together. There were also confusing elements, which have only recently been clarified. For example, it was found that males and females affiliated the most at Wamba, but this included mother-son bonding.

I saw the significance of GG-rubbing most clearly when group compositions were being shifted at the San Diego Zoo. Immediately following these changes, the females were obviously not yet "bonded." They did show an interest in GG-rubbing with one another though. In one group, in particular, the females tried to GG-rub at every occasion. The group male, Vernon, didn't like their behavior. Whenever they GG-rubbed, he would mount a massive display, screaming and jumping over them, sometimes hitting them on the back. The females began to wait, acting as if they had no idea of interacting with each other, and watch him walk off over the hill. Both females would then rise, look to the left and to the right, as if checking where Vernon was, and quickly sneak behind a tree to GG-rub together.

It was impossible for Vernon to be on the alert 24 hours a day. After a couple of weeks, the females did indeed seem to have a bond, and Vernon gave up interfering. I had the feeling that he wanted to prevent them from GG-rubbing because it is part of their bond formation, which was ultimately not in his interest, since females use these bonds to keep males away from food and to attack them.

FdW: How widespread is female dominance in captivity?

PARISH: My best data are from the Stuttgart Zoo where I conducted experiments, the ones in which chimpanzees and bonobos were allowed to fish for honey. In the chimpanzee group, the male always fed first, but in the bonobo group, the females had feeding priority. There was always sex between the females in this situation, especially before they fed together. Sex apparently facilitated their relationship, thus helping them dominate the male.

I have seen similar situations at the Frankfurt Zoo and the San Diego Zoo. At the latter zoo, if sugarcane was bundled, a female would typically grab it. The females would then divide it up amongst themselves, with Vernon, the alpha male, sitting off to the side. There were days on which both adult males just wouldn't get anything. The only exception occurred just after the group merger described above. The females were still unfamiliar and unbonded. Then Vernon did an aggressive display and would take the food for himself.

FdW: You say that females attack males. Do they ever cause injuries?

PARISH: The current alpha female of Stuttgart Zoo is definitely dominant over the adult male. The same applied to the previous alpha female. It is assumed that she once bit his penis almost in half. She was always attacking him, and one day his penis was found cut cleanly in two, hanging by a tiny piece of skin. The vets managed to reattach it.

Frankfurt Zoo has a menopausal female. She's about forty-five years old, the oldest in captivity. She was definitely dominant over anyone else, including her three daughters. The females occasionally held down the male and attacked him, and have bitten off parts of his fingers and toes.

I sometimes wonder whether the high rate of physical abnormalities in males at Wamba may at least partially stem from female aggression.

FDW: Why is female bonding emphasized in bonobo society?

PARISH: There are lots of disadvantages to females that live in a male-philopatric system [a system in which males stay in the natal group and in which inbreeding is prevented by female migration]. In the chimpanzee, this system makes sense, as females need to disperse to solitary ranges to gather enough food. The bonobo habitat, however, is richer and capable of supporting female aggregations; in such an environment, females might be better off staying together with female kin.

So, why did bonobos not revert to a female-philopatric system—that is, a system in which females remain in their natal group? Perhaps it was evolutionarily too costly to completely revamp the existing social organization. The next best alternative for females would then have been to behave with unrelated group-mates as they would with female kin. By mimicking a female-philopatric system, they regained some of its advantages: they in essence remade their own destiny.

While some feminist scholars believe that bonding among women is uniquely human, the extent of bonding among bonobo females probably exceeds that found among women in most human societies.

INCEST AND INFANTICIDE

It may seem thus far as though bonobos are sexually indiscriminate, and that there are no limits. They may, however, be expected to have partner preferences—perhaps even fairly stable ones—and above all to avoid incest, which can lead to inbreeding. Animals generally have inhibitions against inbreeding, since it reduces the viability of their offspring. The bonobos' chief preventative is female migration. When an adolescent female leaves her mother and siblings to join a neighboring community, she goes through a difficult transition, which certainly would not be worth the risk without a significant benefit. The trade-off is that she gets to mate with unrelated males, hence avoids inbreeding.

I should note that in this discussion of bonobo sexuality from an evolutionary perspective, I do not in any way imply that the animals themselves are aware of why they act the way they do. It is safe to assume that we are the only creatures on earth to know the connection between sex and procreation. No one believes that bonobos, or other animals, have any notion of genetics or of the deleterious effects of inbreeding. Migrating bonobos simply follow a tendency produced by natural selection: in the course of the species' history, females who migrated produced healthier offspring than females who did not.

There are no indications that females are chased out of their natal groups or kidnapped by neighboring males. At a certain age, they simply become vagabonds. Kano describes the dramatic change in social relations accompanying the move away from home: "As females approach adolescence, they become less social than males. They occupy the periphery of the group and often sit alone in a tree. This may be a preparation for leaving the group and, occasionally, emigration occurs suddenly after a female has entered the adolescent stage. . . . At some point during the daughter's adolescence, the mother-daughter relationship is completely severed."[14]

When Chie Hashimoto and Takeshi Furuichi studied the sexual behavior of wild bonobos, they found that whereas males become sexually more active with maturation, females do not. On the contrary, the investigators speak of a "sexually inactive" state in prepubertal females—surely a remarkable state to be in for a bonobo. Perhaps it keeps them from developing sexual relations with their brothers and possible fathers. Females generally leave the natal group at the age of seven, when they develop their first little swellings. Equipped with this effective passport, they become "floaters," visiting neighboring communities before permanently settling down in one. All of a sudden, their sexuality flowers. They GG-rub with females and copulate with males encountered in strange forests. They now have regular, almost continuous swellings, which grow in volume with every cycle, until they reach full size at the age of perhaps ten. They can expect their first offspring by the age of thirteen or fourteen.[15]

Possibly, then, a young female's sexuality is suppressed until she needs it for social integration in an environment in which the chances of being made pregnant by her kin are sharply reduced. A delay of several years between her first swellings and actual menarche (known as a period of "adolescent sterility") further protects her against undesirable fertilizations during this stage of her life.[16]

For males, the situation is totally different. They remain in their natal group, and since they cannot get pregnant, they do not risk anything by having sex with relatives. It is the females in their group who stand to lose from such contact. We therefore assume inhibitions against sex with mothers and sisters. The way these inhibitions may come about is through early familiarity—the basic mechanism assumed to underlie incest-avoidance in a wide range of species, including our own. The principle is simple: individuals of the opposite sex with whom one has grown up since infancy fail to arouse sexual feelings. If this process is disrupted—as when zoos raise young apes in a nursery—sex between relatives is not that unusual. Without a common background, there is no way of knowing, so to speak, to whom one might be related. Normally, however, early familiarity characterizes the relations of males with close female relatives, and keeps them from breeding.[17]

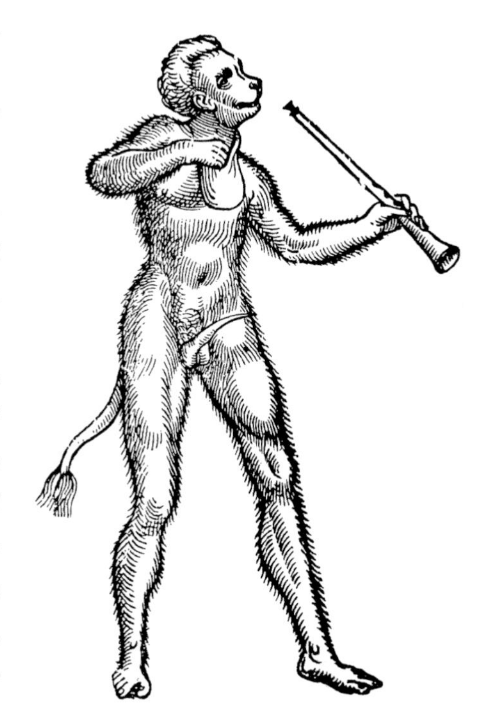

In Western thought, monkeys and apes have historically been associated with lust, rape, and sexual permissiveness. In the seventeenth century, for example, satyr apes on the mythical Satyrides Island, off the coast of India, were said to abuse women. This old gravure depicts such a creature, complete with flute, tail, and bonobolike penis. (From Edward Topsell, The Historie of Foure-footed Beastes *[London, 1607].)*

Hashimoto and Furuichi do report genital contacts between mother and son, but only at an early age and more as a form of self-reassurance than out of sexual motivation: "In all cases involving males younger than two years, mothers held their sons ventro-ventrally and rubbed their genitals against their own genitals. It seemed that mothers performed this behavior to reduce their own emotional arousal, because this behavior occurred only during tense situations, such as when they entered the feeding sites or after they were involved in agonistic interactions."[18]

After the age of two, young males increasingly pursue sexual relations with females, but virtually never with their own mothers. Having recorded only five instances in 137 mother-son units, Kano concludes that incest-avoidance is established at an early age. By the age of four or five, the sexual behavior of young males more and more resembles that of adult males. Swollen females often accommodate the desires of these little Don Juans, who solicit them in the species-typical manner with spread legs and erect penis. They already know how to achieve intromission in various positions.

There is a sharp decline in sexual involvement during a male's adolescence due to the tendency of dominant males to occupy the core of traveling parties, where the females are.[19] Only when they enter adulthood and rise in rank

(continued on page 120)

INFANTICIDE

When a male langur monkey takes over a harem of females after having defeated the original leader, the first thing he does is kill the infants in the troop. He snatches them from their mothers' bellies, mauling them with his sharp canines. In 1967, a Japanese primatologist, Yukimaru Sugiyama, discussed his startling discovery:

> Why does the new leader male bite all the infants? The fact that the new leader male bit all the infants . . . and the fact that many of the females who had lost their infants grew excited and copulated with the new male may be correlated. Because a female langur usually delivers an infant every 2 or 3 years unless she loses her infant, loss of the infant has the effect of advancing the estrus of the female.[20]

With this reflection, the stage was set for an evolutionary explanation of infanticide in primates and other mammals that is debated to this date. The first point of contention is how representative Sugiyama's observations and those of others are. I still remember the pandemonium resulting at scientific gatherings whenever instances of suspected infanticide were reported (often the actual killing remained unobserved: all the audience got to see was postmortem evidence). Differences of opinion about infant-killing were (and are) almost ideological, with one camp willing to accept it as a regular phenomenon and searching for a reason, and another camp viewing it as an aberration: pathological behavior for which there is no room in their vision of nature.

Doubt about whether infanticide is a real phenomenon in wild animals has largely subsided. It is now known to occur in a wide range of species, from lions to prairie dogs, and from mice to gorillas. Current estimates of infanticide as a source of infant mortality (that is, the number of infants that succumb to attacks by conspecifics as a proportion of all infant deaths) are astonishing: 35 percent in grey langurs; 37 percent in mountain gorillas; 43 percent in red howler monkeys; and 29 percent in blue monkeys.[21]

The most widely cited explanation is Sarah Blaffer Hrdy's suggestion that the killing of infants by males is a product of sexual selection. Infanticidal males gain an advantage over other males by eliminating the offspring of their rivals while reducing their own waiting time to fertilize the females whose progeny they have killed. If the genes of infanticidal males spread faster than those of noninfanticidal males, the trait will be favored by natural selection. Naturally, one expects males to target infants unlikely to be their own, such as those carried by strange females. Based on almost twenty years of data from Hanuman langurs around Jodhpur, in India, Volker Sommer has produced the strongest support yet for the idea that infant-killing is a male reproductive strategy.

Infanticide among gorillas and chimpanzees is well known. The very first indication, in the early 1970s, was the observation by Akira Suzuki of a large adult male chimpanzee in Budongo Forest holding a partly eaten dead infant of his own species. Other males were nearby, and the carcass was passed back and forth between them. Dian Fossey saw a lone silverback male gorilla enter a troop with a violent charge. A female who had given birth the night before countered his charge by running up to him, standing upright beating her chest. The newborn on her exposed belly was struck by the male: it died with a wail.

Many more observations and indications of infanticide in wild gorillas and chimpanzees have followed, including an incident that was actually filmed in the Mahale Mountains. Naturally, infanticide is repulsive to human observers, and one fieldworker studying chimpanzees could not resist interfering:

> Mariko Hiraiwa-Hasegawa observed several males surround a female who crawled on the ground and concealed her infant, while she pant-grunted [a submissive vocalization] fervently. Nevertheless, the villainous males attacked her one by one and seized the infant. On seeing this, Hasegawa momentarily forgot her position as a researcher and, brandishing a piece of wood, she intervened and confronted the males to rescue the mother and infant.[22]

If the killing of newborns is indeed absent in bonobos, we need to explain why this species differs from the other African apes, as well as from another close relative, the human species. From King Herod to the child abusers of modern society, threats to the lives of human babies are all too real. How did bonobos escape this curse? Are infanticidal tendencies simply absent in the bonobo male, or did females evolve effective counterstrategies? Perhaps both are true: when females find a way of protecting themselves against infanticide, the tendency may disappear in males.

For the moment we have more questions than answers, but it may well turn out that infanticide holds the key to an evolutionary account of bonobo society.

do males regain access to receptive females. Not that male bonobos are egalitarian with regards to sexual privileges. In contrast to its peaceable image, the species conforms to the general pattern in the animal kingdom of male competition for females. Bonobo males may compete less fiercely than chimpanzee males, but a recent report from Wamba leaves no doubt that dominant males mate more often than others. Since the two top-ranking males in any traveling party generally do most of the mating, it is assumed that they suppress the sexual activity of other males.

Only a tiny fraction of copulations are aggressively disrupted, however. Sexual competition must be rather subtle, therefore, based mainly on fear of what might happen rather than on actual fights. Low-ranking males learn to be secretive about their sexual exploits and develop all sorts of sneaky tactics to attract females. In one case, a male named Mituo waited until the highest-ranking male in the party moved away from Miso, a swollen female. Mituo then climbed the tree in which Miso sat, but passed her and continued climbing until he sat about 4 m above her. After a while, he pulled down an overhead branch, broke off a twig, and dropped it. The twig fell to the ground and was followed by three more, each narrowly missing Miso, who was eating. She must have noticed, but glanced up only once. Mituo swayed his body slightly and dropped several more twigs, all in silence, until—about four minutes after the rain of twigs had started—Miso climbed up the tree and presented herself. Immediately after copulation, she went down to the ground, while Mituo remained in the tree. The high-ranking male had not noticed a thing.[23]

How fitting that the first recorded instance of what might be regarded as tool use by wild bonobos should be related to sex! The fact that low-ranking males need to be careful in the presence of high-ranking ones, and that the latter mate more often, means that sexual competition is more intense than previously thought. It further means that female involvement in male status struggles, as described earlier, may be a way to enhance reproductive success via sons. Natural selection may have favored females who invest energy in their sons' careers if this strategy produced more grandchildren for them.

Given the frequency of copulation among bonobos and the fact that female receptivity stretches well beyond the brief period of ovulation, a male who knew his offspring would have to be a genius. He would need to calculate, perhaps from observed menstruations and births, which females might be fertile, and favor them as sex partners, or at least keep track of when he had intercourse with them. Male bonobos are obviously not doing this; they are just ir-

A unique bonobo behavior is mutual genital contact, or GG-rubbing, between females in a face-to-face position. The female on the bottom clings to the one on top while they rub their swellings sideways against each other. These contacts are brief but intense, serving mostly to alleviate strained relationships.

resistibly attracted to large swellings. That even pregnant females sport these "signs of fertility" makes no difference. Consequently, the bonobo male has no clue which copulations may result in conception and which may not. Almost any little ape growing up in the group might be his, but it could also have been sired by almost any other male, including males from neighboring territories, as a result of intergroup mingling.

If one had to design a social system in which fatherhood remained obscure, one could scarcely do a better job than nature did with bonobo society. Scientists such as Kano, Wrangham, and Parish have begun speculating that this may in fact be the whole purpose of extended receptivity.[24] Females may gain from luring males to engage in frequent sex with almost continuous swellings. Again, no conscious intent is implied, only a systematic misrepresentation of fertility in the service of female reproduction. At first sight, this suggestion is bewildering. What could be wrong with males knowing which offspring they sired? Although paternity is never as certain as maternity, is not our species doing rather well with a relatively high confidence in fatherhood?

True, but before comparing bonobos with ourselves, let us look first at their closest relatives. The shocking truth is that chimpanzee males, as well as the males of quite a few other animal species, are known to brutally kill newborns.

They do not do so often, but often enough for it to pose a grave problem for females. And here I do not just mean the trauma of bereavement. In the cold language of genetic evolution, what matters more is the setback in reproduction and the loss of investment in gestation and lactation. Biologists assume that if a behavior occurs regularly, it must give the performer an edge, otherwise it could never have evolved. What gain could infanticidal males possibly obtain? It is surmised that they force the female to start her reproductive cycle all over again. By eliminating her newborn, males make sure that the female will soon become available for sex and conception. Instead of waiting years for her to resume cycling, the infanticidal male improves his chances at reproduction almost right away.

Thus far, this is the only plausible explanation for infanticide that evolutionary biologists have been able to come up with. It tries to make sense of an act that from almost any other point of view is senseless. The proposed scheme only works, though, if a male can be relatively sure that he himself did not father the infant that he kills. Males prone to make this error would be quite unsuccessful in passing on their genes. To play it safe, therefore, males should chiefly kill the infants of stranger females, or of females known to have consorted with neighboring males. A female who stayed continuously in their territory would be an appropriate target only for those males who failed to fertilize her. Perhaps male chimpanzees follow a rule of thumb that makes them treat the offspring of females with whom they have mated differently from the offspring of females with whom they have never had sex.

Given this male tendency, it is no wonder that chimpanzee females stay away from large gatherings of their species for years after having given birth. Isolation may be their main strategy to prevent infanticide. They resume swellings only towards the end of the nursing period, after three to four years. Until this time, they have nothing to offer males looking for sex, nor an effective way to change the minds of males in an aggressive mood. Female chimpanzees thus spend a large part of their lives traveling alone with dependent offspring.

Bonobo females, in contrast, rejoin their groups right away after having given birth, and copulate within months. They have managed to make paternity so ambiguous that there is little to fear. Bonobo males have no way of knowing which offspring are theirs and which not. Moreover, since bonobo females tend to be dominant, attacking them or their offspring is a risky business. Most likely, if a male were to make a suspicious move, females would band together in defense. We do not know this for certain, because infanticide has thus far never been documented in the species. Perhaps the female counterstrategy is so effective that not even attempts in this direction take place. Unfortunately, it is impossible to prove that behavior *x* does not occur because of reaction *y* that also does not occur, but it is worth speculating along these lines. The relatively carefree existence enjoyed by female bonobos contrasts sharply with the risks faced by female chimpanzees. It is hard to overestimate the premium that evolution must have placed, at least for females, on calling a halt to infanticide.

INTIMATE RELATIONS

An adult male bonobo grooms an adult female (note the prominent breasts). Captive studies indicate that males seek contact with females but that females prefer one another's company—a rather puzzling preference, given that females are also the migratory sex and thus are unrelated to each other. Female bonding is perhaps the sharpest contrast between bonobos and chimpanzees.

A female bonobo climbing a tree (above). The small size of her infant (note the hands and feet) makes it highly unlikely that she is fertile at this point, yet she has a genital swelling. The male (opposite) carries sugarcane while exhibiting an erection, inviting a female for sex in typical bonobo fashion: food triggers sexual excitement, and sex, in turn, paves the way for sharing. O V E R L E A F : *Two bonobo females engage in intimate GG-rubbing while two curious infants approach. Often sexual contacts involve more than two individuals; youngsters, especially, like to get in on the act.*

A B O V E : *Juvenile bonobos get an early start on sex games.* O P P O S I T E : *One female leads another to a quiet spot for some undisturbed GG-rubbing. At the time this photograph was made at San Diego's Wild Animal Park, one male was systematically interfering with intimacies among females. Once the females had gained dominance over him, his interventions ceased, and the females could GG-rub in the open.*

CHAPTER 5

BONOBOS AND US

No single one of the existing manlike apes is among the direct ancestors of the human race; they are all the last scattered remnants of an old catarrhine branch, once numerous, from which the human race has developed as a special branch and in a special direction. *Ernst Haeckel, 1896*

Some human societies have known extraordinary sexual freedom. An extreme case were the peoples of the Pacific before the arrival of Westerners, who brought not only Victorian values but also venereal disease. Bronislaw Malinowski depicted cultures in this region as having few taboos and inhibitions, and the Hawaiians' remarkable sexual permissiveness led the American sexologist Milton Diamond to call sex "a salve and a glue for the total society."[1]

Perhaps there is a bonobo in all of us, as participants to a symposium on human sexuality were once tempted to speculate. Even if people do not engage in sexual rapprochement in the same public manner as bonobos, in the privacy of their homes, similar processes take place. It is not for nothing that the French speak of "reconciliation on the pillow" (*la réconciliation sur l'oreiller*). Our fascination with bonobos is precisely because, consciously or unconsciously, we recognize the way sex functions in their social relationships. There was a time when such issues could not be openly discussed, not even with regard to animals. Animal sexuality was kept at a distance; it reminded us too much of our own carnality. As epitomized by the chimpanzee's old taxo-

A bonobo male makes close eye contact with an infant that could well be his. Given the sexual patterns of this species, however, it is highly unlikely that a male can distinguish his own offspring from that of other males.

nomic name, *Pan satyrus,* monkeys and apes were judged lascivious. Vernon Reynolds attributes this unfavorable opinion, which started with the early Christians, to the moderation of "appetites" that marked the rise of their religion. Self-denial became a source of strength. Any form of gratification, including sexual pleasure, required moral justification. Before this time, monkeys and apes had been seen as naughty and entertaining; now they were seen as despicable:

> Early Christianity taught that sexual behavior, except in the cause of reproduction and within the bounds of marriage, was a sin, and although the idea must have been frustrating to most level-headed pagans of the day, it gained currency with the spread of the new religion and is still widely held in the Christian world today. This frustration found some relief in the aggressiveness of the Christians towards other religious groups, most of which were more tolerant of man's natural desires. And it showed itself in minor ways too, including man's attitude towards the monkey. For those characteristics of the monkey that amused the ancients—its imitativeness, greed, and sexuality—failed to amuse the Christians; these qualities now took on sinister proportions as sins, for which the soul of man would be eternally condemned to hellfire. From being an object of fun and games, the monkey became nothing less than a symbol of the Devil himself.[2]

It appears that people are unable to look at nature without prejudice. We revere animals—such as the busy bee—that confirm the way we would like to be, and we revile animals—such as the gluttonous pig—with urges that we seek to suppress. Science is not entirely neutral either: its attention often ties in with the sociocultural context of the day. Thus, in the postwar years, students of behavior, dismayed by the human capacity for evil, were fascinated by the inborn nature of aggression. And during the revival of free-market ideologies and the decline of communism in the 1970s and 1980s, Neo-Darwinists elevated the pursuit of self-interest to nature's leading principle.

Seen in this light, the bonobo arrives at an interesting turn in history. First and foremost, recent findings seem a belated gift from science to the feminist movement: they provide a concrete alternative to "macho" evolutionary models derived from the behavior of baboons and chimpanzees. Secondly, bonobos thoroughly upset the idea that sex is solely intended for procreation. From now on, any reference to biology in support of this claim will backfire: if something is so obviously untrue of an animal that shares over 98 percent of our genetic makeup, it may very well be untrue of ourselves as well.

Yet even if the psychological mechanisms that allow sexual behavior to be used for purposes other than reproduction derive from a common ancestor, the exact role of sex in human society depends on an evolutionary and cultural history that has been separate from that of our closest relatives for millions of years. The most remarkable product of this evolution is the nuclear family.

FAMILY VALUES

From a female point of view, chimpanzee society seems a rather stressful arrangement. Male chimpanzees do share food with females and are most of the time on good terms with them, but they are supremely dominant, and instead of helping out with offspring, they sometimes pose a threat. Bonobo society offers females a more relaxed existence. Females control the resources, dominate the males, and have little to compete over aside from their sons' careers. The rich forest habitat of the bonobo evidently permits such an organization.

Our ancestors, however, adapted to a much harsher environment. It is dubious that a bonobolike primate could have made it in a savanna habitat while keeping its social system intact. Food is more widely distributed on the plains, which means greater travel distances, especially if many mouths need to be fed. Given visibility to predators, could females and young have ventured far from the forest? Bonobos may be fast, but they are no gazelles: a juvenile bonobo would seem easy prey. Perhaps bands of agile males might have protected the group and helped carry juveniles to safety. But somehow I find it hard to imagine that socially peripheral companions left in the dark about paternity, much like bonobo males, would have been of great help to females.

Male chimpanzees, on the other hand, hunt together, engage in warfare over territory, and enjoy a half-amicable, half-competitive camaraderie. Their cooperative, action-packed existence resembles that of the human males who, in modern society, team up with other males in corporations within which they compete while collectively fighting other corporations. That the same delicate balance between internal strife and unity towards the outside is found in chimpanzee males gives them the most humanlike social system of all the apes. Nevertheless, chimpanzees remain far removed from us when it comes to nurturing the young, which is entirely a female affair. If life on the savanna indeed induces female reliance on males, the chimpanzees success in this habitat would have required a solution to the problem of forced proximity between mothers and potentially infanticidal males. The current chimpanzee social system is adapted to a habitat with dispersed food sources that allows females to forage on their own.

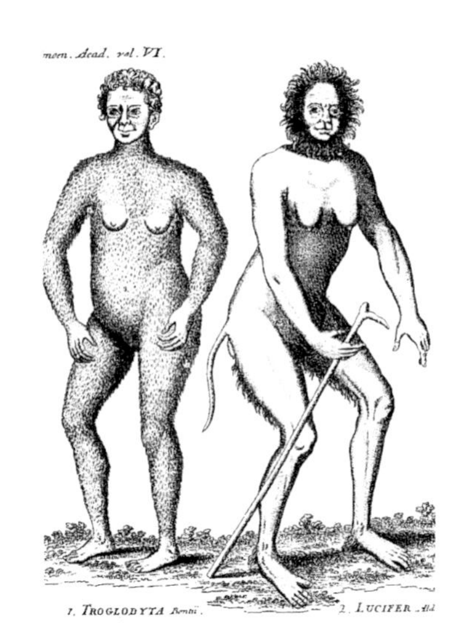

For a long time, scientists had no idea where to draw the line between humans and other hominoids. C. E. Hoppius, a student of the great Swedish naturalist Carl Linnaeus, strongly believed that apes resembled people. In 1760, he published this picture of Anthropomorpha *(creatures formed like humans), based on the limited knowledge of his time. He arranged them from the more humanlike,* Troglodyta *and* Lucifer *(an imaginary race of tailed people), to the more apelike,* Satyrus *(based on the first ape—perhaps a bonobo—dissected by Nicolaas Tulp in 1641) and* Pygmaeus.

So, if we assume that females and young could not have survived on the plains without some degree of male assistance and protection, neither the female-centeredness of the bonobo nor the relatively independent lifestyle of the chimpanzee would have prepared our ancestors for the exploitation of this environment. Bonobolike and chimpanzeelike apes may well have partially ventured into open habitats—as some chimpanzee populations do today—but our ancestors are the only hominoids who managed to abandon the safety of the trees altogether.[3]

Human society is characterized by a combination of (1) male bonding, (2) female bonding, and (3) nuclear families. We share the first feature with chimpanzees, the second with bonobos, and the third is specifically ours. It is no accident that people everywhere fall in love, are sexually jealous, know shame, seek privacy for sexual intercourse, look for father figures in addition to mother figures, and value stable partnerships. Even Malinowski's hedonic "savages" most likely were not without these predispositions, which reflect our inclination to form exclusive households in which both males and females invest in their children. These reproductive units may be monogamous, but the culturally most widespread pattern for our species is polygyny—that is, one male with several females, who may live together or apart. We have been adapted for millions of years to a social order revolving around these nuclear families—the proverbial cornerstones of society—for which no parallel exists in the great apes.[4] This special feature provided our hominid ancestors with a foundation upon which to build cooperative societies to which both sexes contributed, and in which both could feel secure.

Only when males can determine with some certainty which young they have sired do they have any reason to get involved in their care. Perhaps the nuclear family arose out of a male tendency to accompany females with whom they had mated so as to repel infanticidal males.[5] Such a simple security arrangement would have been easy to expand upon. For example, the father could have helped his companion locate fruit trees, capture prey, or transport juveniles. He himself might have benefited from her talent for precision tool use and the gathering of nuts and berries.[6] The female, in turn, may have begun prolonging her sexual receptivity so as to keep her protector from abandoning her for every good-looking passerby. The more both parties committed to this arrangement, the higher the stakes. It became increasingly important for the male that his mate's offspring were his and only his. From an evolutionary perspective, investment in someone else's progeny is a total waste of energy, hence males tightened control over their mate's reproduction in direct proportion to the assistance they gave her.

In nature, nothing is free. If bonobo females paid for their successful arrangement with almost continuously swollen genitals, hominid females paid for theirs with the loss of reproductive freedom. The reasons for male control only doubled when our ancestors settled down from a nomadic existence and began to accumulate material goods. In addition to passing on genes to the next generation, there now was also the inheritance of wealth. Since male dominance has probably always characterized our lineage, inheritance tended to take place along paternal lines. With every male trying to ensure that his life savings ended up in the right hands—that is, the hands of his progeny—an obsession with virginity and chastity was inevitable. Without male involvement in the raising of offspring, there would be no need for the moral constraints, sometimes referred to as "patriarchy," that our species universally employs to safeguard the integrity of the family.

Talk about extremes! The chimpanzee male is keen to establish which infants are *not* his, so as to terminate their lives, whereas the human male has evolved a paternal investment strategy requiring that he do everything he can to ensure that he is himself the father of his mate's offspring. The nuclear family increases confidence in fatherhood.

And the bonobo? According to these theories, the bonobo has chosen a third way by confusing the whole issue: if all males are potential fathers, none of them has a reason to harm newborns.

BONOBO SCENARIOS

Is evolutionary modeling an art or a science? Compared to the way a paleontologist pieces together the human past, a primatologist trying the same for a particular species of anthropoid ape has little concrete material to go by. Our immediate ancestors at least left artifacts behind that allow us to reconstruct their habits, skills, means of subsistence, and social organization. Primatologists, on the other hand, have to content themselves with the behavior of extant creatures.

Despite the undeniable role of imagination and speculation, however, designing a plausible scenario of a species' evolution is a scientific endeavor—that is, its outcome is open to falsification. Each scenario revolves around a set of assumptions that can be proven wrong. A good example is the way the recent discovery of "Little Foot" (see p. 26 above) contradicts existing views about the origin of human bipedality. Similarly, theories about bonobo social evolution would be in deep trouble if it were found that male bonobos are perfectly capable of discriminating between a female's fertile and infertile days. Such observations have not yet been and may never be made; the point is that evolutionary models are subject to verification like all other scientific hypotheses. Whereas it is impossible to demonstrate the correctness of any particular model of the past, our goal is to select the scenario that agrees best with the available data.

The defining moment in the social evolution of the bonobo probably took place when females began to have more frequent and longer-lasting genital swellings. Together with a general sexualization of the species, this reduced competition among males, obscured paternity, and promoted sociosexual relations in all partner combinations, particularly among the females. The end result was that females formed a secondary sisterhood, gained the upper hand in society, and freed themselves from the curse of infanticide.

The initial stages of this evolutionary scenario remain the most obscure. The so-called "sexualization" of the bonobo—meaning that sexual behavior began to permeate all aspects of social life—most likely started between the sexes. After all, the original function of sex is reproduction, which implies adult heterosexual relationships. Perhaps both sexes gained from each other's company and from tolerant and friendly relations promoted by frequent sex. Since female mammals almost never dominate adult males of their species, it is safe to assume that bonobos started out with male dominance. Sex-for-food transactions may have helped females gain access to food controlled by males. Consequently, females gained from the extension of sexual receptivity. If male-female associations were nonexclusive—that is, if they involved multiple males and females—tolerance between members of the same sex became an issue as well. Conceivably, the role of sex as the cement of society spread from the heterosexual domain to other domains. Homosexual activity became a way of tying males and females together in larger aggregations.[7]

The question of why bonobos and not chimpanzees sexualized their social relations is hard to answer, however, without reference to infanticide. Sexual behavior is by no means required for bonding and tolerance. Chimpanzees, for example, employ nonsexual behavior that is as effective as the genital contacts of bonobos. Chimpanzees reconcile with a kiss and share food after "celebrations" marked by loud vocalization and body contact in which they hug and pat each other on the back. The reliance of bonobos on sexual mechanisms may be because of the added benefit of obscuring pater-

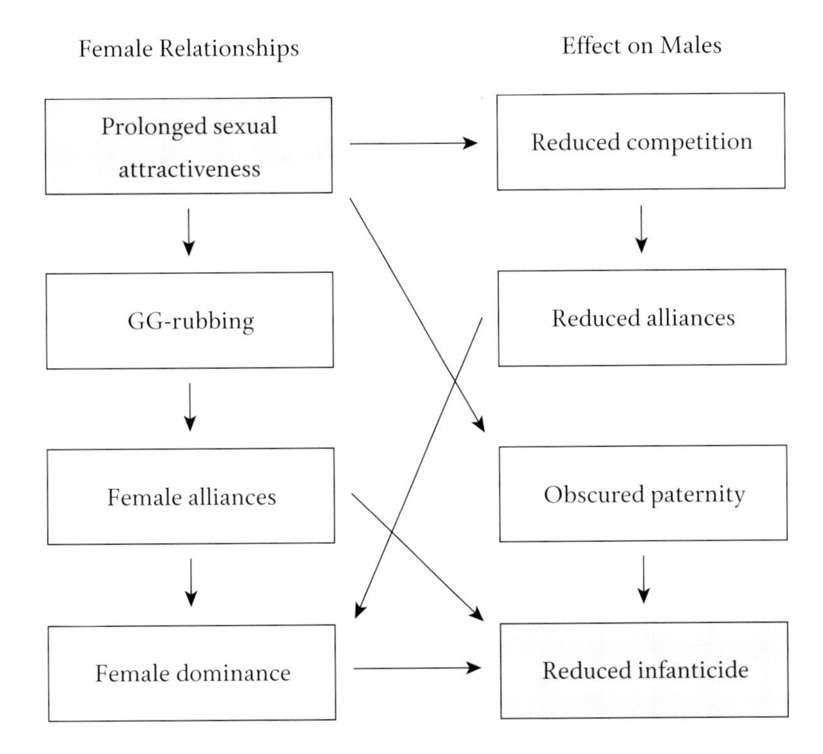

nity, especially in a species in which males and females tend to travel and forage together.

I must confess to being uncomfortable with hanging an entire evolutionary scenario on the menace of infanticide, which has never been reported for bonobos, captive or wild. We also need to consider what, if anything, male bonobos gained from these changes. If females improve their reproductive success, some or all males improve theirs as well: males procreate through females.[8] This poses an interesting challenge for scientists trying mathematically to simulate the evolutionary consequences of social arrangements. How do male interests fare under rising female influence? Would it be worthwhile for males to rebel, or are some or all of them actually better off "playing along"? Moreover, the unique structure of bonobo society may carry advantages that we at present cannot even fathom. We know little about predation pressure, past or present, or about competition with other fruit-eaters in the forest. The scenario depicted in the above scheme must be regarded as tentative, therefore. But even if not accepted in its entirety, it has plausible subcomponents. The central ones are

1. Extended female receptivity dilutes competition among males. The presence of so many attractive females, and the impossibility of pinpointing their fertile days, makes it less worthwhile for males to risk injury over a mating.

2. Given that male alliances in other primates are mostly instruments to keep competitors away from a highly contested female, the reason for such cooperation is eliminated if multiple females are sexually attractive at once.

3. An every-male-for-himself system paves the way for a collective female power takeover.

4. Sociosexual behavior and bonding among females translates into alliances that allow them to monopolize food and protect their offspring against infanticidal males.

5. Extended receptivity and frequent sex confuse paternity to such a degree that infanticide becomes counterproductive: males have trouble exempting progeny.

Each part of this scenario deserves close scrutiny by scientists intent on reconstructing the bonobo's evolutionary past. We need to be especially keen on possible exceptions. If, for example, a bonobo population with occasional infanticide were found, or if female dominance turned out to be less widespread than currently thought, this would pose major problems. Even though a few exceptions are unlikely to torpedo the entire model, it will probably require revisions in the years to come.

The proposed scenario also raises intriguing questions. I have heard men wonder aloud, in an almost accusatory tone, "Why did these bonobo males let the females take over?" Is this the right way of putting it, though? The word *let* presumes that bonobo males had a choice, but what if they didn't? Or take the question of why, if the bonobo social system is so advantageous, chimpanzees did not go down the same evolutionary path? No doubt, ecological conditions were a key factor. Without large enough food concentrations to permit associations of multiple males and females, the entire scheme could never have worked. How could female solidarity ever have been achieved—and how could it have operated—if bonobo communities had been forced to split into small parties in their quest for food?

That male bonobos do not band together to put a halt to female ambitions is perhaps because the payoffs of being dominant are not substantial enough to achieve effective cooperation. Males are always in competition about mating rights, which makes alliance formation a tricky process. It hinges on what each party to the alliance, not just the eventual winner, gets out of it. Few animals are capable of striking mutually profitable deals. Chimpanzee males are a notable exception: dominant males selectively share food and sexual privileges with male allies to whom they owe high status. There is much less to share in a species, such as the bonobo, in which males seldom, if ever, hunt for meat, and in which receptive females are relatively plentiful. What would be

With an infant under her belly and a juvenile on her back, this bonobo mother bears the burden of a single-parent system with reduced intervals between births.

the point, for example, of two or more males guarding a particular female if other receptive females are available to their competitors? While it may pay for males to compete individually over females, the advantage of monopolizing a female may never reach the threshold at which alliance formation becomes an attractive strategy.[9]

Apart from the lack of a sound basis for male cooperation, there is the rising influence of the mother. The higher the status of females, the more mothers could do to help their sons in their quest for high rank. When maternal support became as effective for males as support from other males, this further undermined whatever tendencies toward male cooperation may have existed. It became increasingly advantageous to rely on a less fickle partner, such as the mother. Eventually, the nucleus of traveling parties shifted from association between members of the opposite sex to association between mother and son, as well as the powerful alliance among senior resident females. At this point, sex-for-food exchanges between the sexes were not needed anymore. These exchanges may be a vestige of a past in which females had not yet attained dominance; they are typically initiated by young females.

Did our social system derive from one similar to the bonobo's? Taking a bonobolike ancestor as starting point, human evolution would have required dramatic changes, such as the loss of female genital swellings while retaining prolonged receptivity, an overturn of female dominance, the evolution of male cooperation in hunting and warfare, and establishment of the nuclear family. How and why these changes should have taken place is unclear; it seems more economical to assume that the common ancestor of apes and humans had no swellings at all and knew both male dominance and male cooperation. With regards to the latter aspect, fewer changes might have been required to get from a chimpanzeelike social system to ours: this species at least knows aggressive male cooperation. However, a family-based society seems even remoter from the chimpanzee than from the bonobo.

For the moment, the safest assumption is that all three species—humans, chimpanzees, and bonobos—are specialized—that is, that they have evolved considerably since their descent from the common ancestor. Unfortunately, this leaves unresolved the puzzle of the missing link's social life: no extant species can be adopted as model. The good news, however, is that in the bonobo we now have an additional key to unlock the mystery: this species may have retained different traits from the ancestor than the other two. The bonobo shows an unparalleled social organization that should give pause to anyone claiming the universality of certain traits in our ancestry. If two close relatives such as bonobos and chimpanzees differ so greatly, this hints at a flexibility in our lineage—not just in the cultural sense, but also evolutionarily—that many of us had not held possible.

In all of this, we should keep in mind that the bonobo and the chimpanzee are equidistant from us. Rather than favoring parallels with one or the other ape, there is no need to choose between the two. Those who for ideological reasons are inclined to advocate the bonobo as the model of the missing link should realize that evolutionary biology does not permit such selective attention. One cannot nibble on a little piece of it without swallowing the entire pie. The bonobo as ancestral model comes with an entire framework of evolutionary thought. This framework tries to accommodate the behavior of baboons, gorillas, chimpanzees, and a host of other species. Most biologists consider the general principles of adaptation and natural selection of greater importance than the evolution of a particular species. It is true that the bonobo, by virtue of its close relation to us, is a critical piece in the puzzle of human evolution, but it is really the entire puzzle that science seeks to solve.

Clearly, the most successful reconstruction of our past will be based on a broad, triangular comparison of chimpanzees, bonobos, and ourselves within this larger evolutionary context.

MAKING SENSE

The vocal repertoire of the bonobo is unique and very well developed. The bonobo's voice is so much higher than the chimpanzee's that the easiest way to tell the two species apart is by ear. In dense forest, vocalizations are often the sole means of communication; visual signals are of use only at short range. Calls are used to bring individuals together or to drive them apart. Most likely, a bonobo can recognize the voice of each individual in its group.

Charles Darwin remarked on the similarities in facial expressions between humans and other primates. The nervous grin of this young male (above, left) and the pout of an infant (above, right) are typical bonobo expressions that convey the individual's emotions. Other expressions are more deliberate and less emotionally charged: this captive female (opposite) opened her mouth in a protracted yawn to provoke a reaction from the photographer.

Bonobo males may play a decisive role in group movement. An adult male (opposite) runs into the forest while dragging a sizable branch behind him. This brief display, which generates a great deal of noise, serves to attract attention and initiate movement. Ellen Ingmanson, an American anthropologist who recorded hundreds of such displays at Wamba, found that they predicted the subsequent travel direction of parties. Performers of these displays appear to "suggest" to their companions where to go—perhaps to a fruiting tree that they know about or a stream where they regularly forage.

In addition to facial and vocal communication, apes often gesture at each other. Here we see an exchange between two young bonobos before they play. Bonobos tend to gesture with their right hands, which hints at a brain specialization similar to the one underlying human language.

CHAPTER 6

SENSITIVITY

When I saw bonobos for the first time at the Wassenaar Zoo, in 1978, I was anxious to see how they compared to chimpanzees.[1] What struck me was not only how much larger bonobos were than suggested by the then-common name "pygmy chimpanzee," but especially the inquisitiveness and sensitiveness of their gaze. Such lively eyes, and so openly interested in the people around them, including a stranger such as myself!

At the time, virtually nothing was known about the species' social life. Had we known then what we know today, the bonobo would no doubt have become a thorn in the side of science. Perhaps they would have been declared an "abnormality," to be disregarded by anyone interested in the general principles of life on earth. After all, in those days we saw ourselves as nasty killer apes, endowed with an irrepressible urge to butcher one another. The chimpanzee was the perfect primate model for this time—especially after the discovery of intraspecific killing and cannibalism in the species. The bonobo would not have fit.

A "babysitting" female plays with another female's infant that is swinging from a tree.

We now live in a different time. Most of us have heard quite enough about the aggressive tendencies of humanity. Of course, we recognize their existence—one only needs to watch the daily news to see the most gruesome examples—and we recognize their inborn origins, at least most biologists do. Yet at the same time our attention has shifted towards mechanisms that keep these tendencies in check, and that allow for peaceful coexistence. Living in a relatively nonviolent, egalitarian, and female-centered society, the bonobo may have much to offer in this regard. Competition is by no means absent in the species, but by and large bonobos manage social discord remarkably effectively.

My previous speculations about the origins of bonobo society represent only the evolutionary side of the story. The animals themselves know nothing about how evolution has shaped their behavior. Bonobos are not calculating the reproductive consequences of their behavior or the advantages of obscured paternity. Females cannot make genital swellings appear or disappear at will, anymore than males can consciously weigh the costs of infanticide. Inborn characteristics, including behavioral ones, that promote reproduction are handed down from one generation to the next without any conscious decision on the part of the animals themselves. It is one thing to conjecture why a particular behavior may have been favored by natural selection but an entirely different problem to understand how animals live their daily lives.

The day-to-day decisions of apes revolve around avoiding or reducing tensions, seeking and giving pleasure, maintaining bonds, protecting the young, getting enough to eat, and so on. Bonobos are born with a set of psychological, physiological,[2] and social predispositions that dictate what kind of society they will build. A key predisposition may be the sensitivity that I noticed during my first encounter: it may provide a foundation for what in humans we call "sympathy" and "empathy."

Understanding the intentions and feelings of others may help bonobos smooth relationships, provide assistance where needed, and intensify sexual experiences. Conflict resolution, for example, depends on taking early notice when something bothers a companion, and on knowing what to do so as to prevent frustration. In the sexual domain, we have seen suggestions that bonobos regulate their performance based on what they read in their partner's eyes. They also commonly stimulate each other without stimulating themselves,

(continued on page 158)

Thumb-sucking is a universal primate way to obtain oral stimulation when the mother has begun the weaning process.

SENSITIVITY TO OTHERS

Increasingly, scientists are exploring whether primates aside from ourselves are able to understand the intentions, thoughts, and feelings of others. Can they mentally put themselves in someone else's place? Do they recognize the needs and wants of others? Here are a few anecdotes suggesting that bonobos possess the precious capacity of cognitive empathy.

BIRDS SHOULD FLY

Betty Walsh, a seasoned animal caretaker, observed the following incident involving a seven-year-old female bonobo named Kuni at Twycross Zoo in England.

One day, Kuni captured a starling. Out of fear that she might molest the stunned bird, which appeared undamaged, the keeper urged the ape to let it go. Perhaps because of this encouragement, Kuni took the bird outside and gently set it onto its feet, the right way up, where it stayed, looking petrified. When it didn't move, Kuni threw it a little way, but it just fluttered. Not satisfied, Kuni picked up the starling with one hand and climbed to the highest point of the highest tree, where she wrapped her legs around the trunk, so that she had both hands free to hold the bird. She then carefully unfolded its wings and spread them wide open, one wing in each hand, before throwing the bird as hard she could towards the barrier of the enclosure. Unfortunately, it fell short and landed onto the bank of the moat, where Kuni guarded it for a long time against a curious juvenile. By the end of the day, the bird was gone without a trace or feather. It is assumed that, recovered from its shock, it had flown away.

GETTING SOMEONE A DRINK

Thomas Patterson observed the following recognition of another individual's desires in the San Diego bonobo colony in 1971.

Linda's two-year-old daughter uttered whimpering sounds, looking at her mother with a pouted expression. Normally, this means that an infant wants to get nursed, but all of Linda's offspring were human-reared. The infant had been returned to the group long after Linda had stopped lactating. The mother went to the fountain to suck water into her mouth. She then sat down in front of her daughter puckering her lips so that the infant could drink from them. Linda repeated her trip to the fountain three times.

A HELPING HAND

Kidogo, a twenty-one-year-old bonobo at the Milwaukee County Zoo suffers from a serious heart condition. He is feeble, lacking the normal stamina and self-confidence of a grown male. When first moved to Milwaukee Zoo, the keepers' shifting commands in the unfamiliar building thoroughly confused him. He failed to understand where to go when people urged him to move from one place to another.

Other apes in the group would step in, however, approach Kidogo, take him by the hand, and lead him in the right direction. Barbara Bell, a caretaker and animal trainer, observed many instances of such spontaneous assistance and learned to call upon other bonobos to move Kidogo. If lost, Kidogo would utter distress calls, whereupon others would calm him down or act as his guides. One of his main helpers was the highest-ranking male, Lody. These observations of bonobo males walking hand-in-hand dispel the notion that they are unsupportive of each other.*

Only one bonobo tried to take advantage of Kidogo's condition. Murph, a five-year-old male, often teased Kidogo, who lacked the assertiveness to stop the youngster. Lody sometimes interfered, however, by grabbing the juvenile by an ankle when he was about to start his annoying games, or by going over to Kidogo to put a protective arm around him.

SISTERHOOD

In the course of her studies, Amy Parish developed close relations with zoo bonobos, and the females treated her almost as one of their own. On one occasion when the San Diego bonobos were given hearts of celery, which were claimed by the females, Parish gestured to have the apes look her way for a photograph. Louise, who had most of the food, probably thought that she was begging and ignored her for about ten minutes. Then she suddenly stood up, divided her celery, and threw half of it across the moat to this woman who so desperately wanted her attention.

Parish had no trouble obtaining fecal samples for laboratory analyses. One day, she simply begged from a female who was holding excrement in

*Nevertheless, they fail to establish effective alliances: the three adult males at the Milwaukee Zoo—two healthy males and Kidogo—are said to be firmly dominated by an older adult female.

her hand. By the end of the week, and without any rewards, all four females were voluntarily handing in samples. As soon as they saw Parish, they began to search for them. When Parish later went to work at the Wilhelma Zoo, in Stuttgart, Lina, the daughter of a female who had been part of the San Diego group when she worked there, remembered her peculiar interest. Even though Parish had never collected from Lina herself, the young female handed her a smelly present.

Another time, when Parish visited the Stuttgart bonobos after a long absence, she showed them her son, who had been born in the meantime. The dominant female briefly glanced at the baby, then suddenly disappeared into an adjacent cage to return with her own newborn.

---

such as when one individual massages the genitals of another. Does not such behavior suggest that they realize what gives the other pleasure?

Could it be that the bonobo is cognitively specialized to read emotions and to take the point of view of others? In short, is the bonobo the most empathic ape? If so, a focus on tangible manifestations of intelligence, such as tool use, might be misleading.

If used at all, tools may serve mostly in the social domain, rather than, for instance, for food acquisition. An example is the use of so-called *taboo nests* observed by Barbara Fruth and Gottfried Hohmann in Lomako Forest. Bonobos build nests in the trees for the night, but also for resting, grooming, or play in the daytime. It is as if these nests represent a private area that cannot be infringed upon, not even by the nest maker's closest companions. For example, youngsters do not enter their mother's nest uninvited, but wait at the edge, requesting access by means of facial expressions and distress calls.

If nests demarcate "personal space," the sharing of which is up to the builder, this allows females to wean offspring and force them to nest elsewhere when they have reached the age to do so. It also allows the avoidance of conflict through the use of nests as a refuge. Fruth and Hohmann documented a dozen cases in which bonobos feeding on favorite foods responded to the approach of a companion by quickly breaking off a few branches and constructing a rudimentary nest. While sitting in the nest, they were not bothered or displaced by the others, and could consume their food undisturbed. One case did not involve food: an adult male escaped the charging display of another male by climbing up a tree and building a nest. In response, the charging male stopped at the base of the tree and moved away.

How bonobos respond to the feelings and needs of others may prove a most exciting area of research. Take Kanzi's striking grasp of spoken English. Rather than analyzing this in terms of linguistic skills, it may actually be better understood in terms of social cognition. Comprehension may rest on the ability to figure out the intent behind the sounds people utter. Kanzi also seems to realize that some of his fellow bonobos do not have the same background: he occasionally takes it upon himself to teach the illiterate. In one striking video segment, Kanzi sits next to Tamuli, a younger sister who has had minimal exposure to human speech. Sue Savage-Rumbaugh tries to get Tamuli to respond to a few simple verbal requests, but the untrained bonobo is at a total loss. Whereas the investigator clearly addresses Tamuli, it is her big brother who acts out the meaning of the requests. At one point, when Tamuli is asked to groom Kanzi, he takes her hand in his, and places it under his chin, squeezing it between his chin and chest. In this position he stares into Tamuli's eyes with what looks like a questioning gaze. When Kanzi repeats these actions, the young female rests her fingers on his chest as if hesitating over what to do.[3]

It should be added that Kanzi knows perfectly well when commands are intended for him or for one of his fellow apes. He was not merely carrying out a command intended for Tamuli, but actually took Tamuli's hand to get her to act upon himself. Kanzi's sensitivity to his sister's lack of knowledge and kindness in pointing things out to her suggest a high level of empathy. Further anecdotal evidence for this capacity can be found in previous chapters (such as Kakowet's reaction when the zoo caretakers were about to fill a moat in which youngsters were playing); but the most astonishing case is perhaps that of Kuni and the bird at Twycross Zoo (see p. 156). To identify with a creature so totally unlike oneself and place its body in a posture in which it commonly manifests itself is, to my mind, as impressive as any kind of tool use.

Unfortunately, no experiments on the ability to take another individual's perspective have been conducted with bonobos. Such studies are increasingly carried out with chimpanzees and young children; adding bonobos as test subjects might produce startling insights. Over seventy years ago, Robert Yerkes was so struck by the concerned attitude of his young bonobo, Prince Chim, towards his sickly chimpanzee companion, Panzee, that he noted: "If I were to tell of his altruistic and obviously sympathetic behavior towards Panzee I should be suspected of idealizing an ape."[4] Thus, even before bonobos were distinguished as a species, their chief mental specialty may already have been recognized by one of the great experts in ape temperament.

Discovery of a close relative that sheds a completely fresh light on human evolution is to be celebrated as one of the greatest strokes of luck to have befallen anthropology and primatology during this century. The bonobo is overthrowing established notions about where we came from and what our behavioral potential is. Without this ape, traditional evolutionary scenarios emphasizing human aggressivity, hunting, and warfare would no doubt have continued to dominate the discussion, despite the fact that our species possesses a multitude of other defining characteristics relating to language, culture, morality, and family structure. Even though the bonobo is not our ancestor, but perhaps a rather specialized relative, its female-centered, nonbelligerent society is putting question marks all over the hypothesized evolutionary map of our species.

Who could have imagined a close relative of ours in which female alliances intimidate males, sexual behavior is as rich as ours, different groups do not fight but mingle, mothers take on a central role, and the greatest intellectual achievement is not tool use but sensitivity to others? Any scientist proposing such a list of traits as even remotely likely in a member of our immediate lineage would have been laughed out of the halls of academe in the 1960s! Today, more people may be prepared to accept this description—all the more so since we are not talking about a mere supposition but about actual observations. Nonetheless, it will undoubtedly take an entire generation of students of human evolution for the implications fully to sink in.

Let us hope that in the meantime the species will survive in the wild, and that fieldwork will continue. Similarly, we must hope that the small captive population will turn out to be reproductively viable, so that we shall be permitted to probe deeper and deeper into this ape's cognition and behavior.

Study of the bonobo has only just begun.

SOCIAL LIFE

Bonobos are highly social creatures, occasionally aggregating in large parties that bring both competitors and friends together. Perhaps this tendency is what led them to develop such remarkably complex and subtle communication. Their sensitivity to the needs of others is borne out in anecdotal reports as well as in the mutual bodily stimulation that these apes frequently engage in. Here an adult female lifts a juvenile's head to better look into her eyes.

Food-sharing is not widespread in the primate order. The few species that do share, including bonobos, use special calls to announce food and gestures to solicit it. One adult female reaches for the food of another (opposite), while an infant begs its mother to drop into its hand a morsel of the food she is chewing (above).

It is often argued that human babies have more eye contact with their mothers than do infants of other species. While this is probably true, ape mothers often play games with their offspring that promote eye contact. A favorite among captive apes, who spend much time on the ground, is the "airplane" game.

In addition to the bonding among unrelated females, which characterizes bonobo society as a "secondary sisterhood," another crucial bond stands out: mother and son maintain close ties throughout their lifetimes. Sons and daughters both start life fully dependent on their mothers (opposite), but whereas a daughter begins to distance herself from her mother at puberty, a son remains attached and continues to enjoy his mother's attention (above).

EPILOGUE

BONOBOS TODAY AND TOMORROW

The reason many readers might not have heard of bonobos before picking up this book is that these apes live in an extremely remote part of the world and few zoos have them on display. Compared to thousands of chimpanzees in captivity, there are only about a hundred bonobos. As for the number of wild bonobos, it is safe to assume that the species is endangered, but it is unclear at this point to what degree this is the case. Perhaps it is still early enough to avert the fate that threatens a host of other animals: the bonobo has the good fortune to live in a relatively undisturbed habitat.

WHERE THE BONOBO DWELLS

Bonobos live in one of the most sparsely populated, least developed parts of the tropics. The Cuvette Centrale region of the province of Equateur, in Zaire, is part of the second-largest solid stretch of rain forest in the world, representing half of all remaining African rain forest. The region is so hard to reach that the most reliable means of travel there is on foot or by dugout canoe. Isolated pockets of bonobos have been located throughout this vast primeval forest.

At the few zoos that have bonobos, visitors can watch these primates up close. The patient observer will be able to follow bouts of social activity interspersed with rest and grooming, as here at the Cincinnati Zoo.

The main road to Wamba is barely passable with four-wheel-drive vehicles, and even bicycles are a luxury. The inaccessibility of the region is a serious detriment to the local economy but perhaps a blessing for the bonobos.

The species' range is bounded to the north and west by the mighty Zaire River, to the east by the Lomami River, and to the south by the Kasai and Sankuru rivers. The area of potential distribution has been estimated at 840,400 km^2, but the species may actually be found in only a quarter of that area.

In the northern half of their range, bonobos perhaps occur in relatively good-sized populations, whereas the distribution may be more fragmentary in the southern half. Both main research sites, Lomako and Wamba, are located in the northern part. Other sites are Lilungu, Yalosidi, Yasa, and Lake Tumba. The southernmost site, Yasa, where an American student, Jo Thompson, began studies in 1994, belies the belief that bonobos can only be found in continuous evergreen forest. In this part of the bonobo range, the habitat alternates between forest, open woodlands, and grasslands. Study of bonobos in the hilly mosaic of forest and grassland around Yasa is bound to provide insights into this ape's ecological flexibility.

There are no reliable estimates of the size of the bonobo population in Zaire. Early figures exceeded 100,000 individuals, but a number between 10,000 and 25,000 seems considerably more realistic. Bonobos deserve their current classification as vulnerable to extinction by the International Union for Conservation of Nature (IUCN), as endangered by the United States Fish and Wildlife Service, and as Class A of the African Convention and Appendix I of the Convention on International Trade in Endangered Species (CITES), which bans hunting, killing, trapping, and trade.

At present, Zaire has only one forest reserve at the national level to protect bonobos: Salonga National Park. There are confirmed sightings of bonobos in this huge park (36,560 km^2), which was established in 1970. Bonobos occur throughout Salonga, but probably never in high densities. The park is presently under the control of heavily armed poacher gangs, which have decimated populations of elephants and hippos. The impact of these gangs on other wildlife is unclear; large parts of the park remain undisturbed.

A top priority for any conservation program will be a detailed survey of existing populations, their distribution and abundance, in the entire range of the species. Such a large-scale survey has never been conducted—and would be a major enterprise—but it will be hard to decide on the most effective conservation measures without it. As explained below, plans for reserves are being developed, but apart from Salonga, there are at present no further reserves to protect the species.

CRUMBLING TABOOS

Wamba Forest used to have a high population density of approximately 300 bonobos in 70 km^2. This forest is part of a somewhat larger protection area, the Luo Scientific Reserve, which also includes a thousand resident people, who coexist with the bonobos. Hunting is prohibited around Wamba according to a document signed by the researchers and local bureaucrats in 1987. This used to work well because of a strict local taboo against eating the meat of bonobos based on ancient beliefs that bonobos are our kin, almost like ancestors.

This taboo was well respected until 1984. In that year, the first poaching incident occurred during an absence of the investigators. A hunter from outside Wamba killed a young adult male, then carried the carcass to a nearby village to sell the meat. The second case, in 1987, was more serious. During another absence of the researchers, soldiers were ordered to Wamba to capture apes. Trackers who refused to cooperate were beaten. During a mass hunting, soldiers killed several adult males and two bonobo mothers to collect their infants. If Kano's chief tracker had not placed himself between the soldiers and their targets, they might have killed even more apes. The rumor is that the two baby bonobos were taken to the capital of Kinshasa to be presented to a national guest.

If we look at the number of individuals who have disappeared from Wamba's main study group, it becomes clear how much things have changed for the worse. During the eight years from 1976 through 1983, only three individuals were lost, but during a period of the same duration, from 1984 through 1991, no fewer than eleven bonobos disappeared. In the past couple of years, another ten apes have been lost. Almost all disappearances occurred during absences of the researchers, and surprisingly many of the lost apes were adult males in their prime (males have a tendency to place themselves closer to hunters). Whereas the group had been growing in size from the mid 1970s through the mid 1980s, it is now declining.

A bucket filled with bonobo pieces, carried by a man from a nearby village, and a dried wrist and fragments of the long bones of a bonobo in a house in another village gave direct evidence of poaching at Wamba. In a survey in a nearby hamlet, 20 percent of the men said they had eaten bonobo. All of them used to observe the taboo against this food until ten years ago. Another bad piece of news is that people have again begun making poisoned arrows—a once obsolete practice—probably because it allows for hunting undetected by researchers and authorities.

The decline in the bonobo population has prompted Kano and his colleagues to start developing a plan for a bonobo sanctuary. The Luo Special Protection Area would be a 6,000 km^2 reserve around Wamba, containing some 50 villages. If this project gets off the ground, it would include a public relations campaign, educational programs, antipoaching patrols, and a survey of the bonobo population in the eastern part of their range. It might help put a halt to the expansion of coffee plantations. The protection area would stretch far beyond the existing Luo Scientific Reserve. But, as can be seen from current problems at Wamba, even if governmental support could be secured for this plan, enforcement of a hunting ban will be a major challenge.

Although there are still stories in the Western media that apes are being captured or killed on a large scale for biomedical research, the truth is that because of successful breeding programs and reduced use of primates, particularly apes, most biomedical facilities now have a surplus of them. In addition, trade has virtually come to a halt thanks to international laws. Now the gravest threats to tropical primates, including bonobos, are habitat destruction and the bushmeat trade.

Industrial exploitation of wood is less of a problem in Zaire than elsewhere in tropical regions because of the inaccessibility of the Cuvette Centrale, but it is nonetheless estimated that 200,000 ha of forest are lost every year to clear-cutting for agriculture, and another 200,000 ha to cutting for firewood and charcoal-making.

A large German veneer manufacturer set up a logging operation around Lomako Forest in 1981, but abandoned the concession in 1987, leaving behind an area of 3,800 km^2 that could be turned into a reserve. Lumber roads to the west, however, have made this area partially accessible to hunters and slash-and-burn agriculture. Plans for a Lomako Reserve were submitted to the authorities in 1990; governmental approval is still awaited.

Apart from charcoal, another "forest product" sold to urban populations is the meat of wild forest animals. Bushmeat brings much-needed cash to people living inland, many of whom lack the barest means of subsistence. At a new bonobo study site, in Lilungu, guns were found to be rare, and bonobos were thought to be protected by local religious beliefs. However, Spanish primatologists discovered more than a thousand wire snares in the forest. Even if not intended for apes, such poaching methods pose a serious threat to bonobo survival, causing injuries that may lead to infection and death.

The saddest by-products of the bushmeat trade are surviving primate infants removed from their mothers' bodies. The young primates are sold as pets, which often means that they die a slow death from malnourishment and inadequate living conditions. Some young bonobos have had the good luck to end up with Delfi Messinger, an American associated with a French biomedical laboratory in the capital, Kinshasa. Messinger has thus far adopted a dozen sick and unwanted bonobo orphans, for whom she and her staff care under the most difficult circumstances. The story goes that when the Belgian and French governments were evacuating 15,000 expatriates because of riots, looting, and plundering in Kinshasa in 1991, Messinger chose to stay and defend her bonobos. She painted the letters "SIDA"—the French equivalent of "AIDS"—across the entrance of the facility. It proved effective in scaring off the looters.

Wild animals, including monkeys and apes, are increasingly being killed for meat in Zaire. Here Wamba villagers cut up a duiker.

BONOBOS AT THE ZOO

Bonobos live long and healthy lives in captivity, and they are not difficult to breed. The problem is that there are altogether only around a hundred of these apes in the world's zoos. So small a population makes it virtually impossible to maintain genetic diversity. Certainly, no single institution can set up a viable breeding program, and even if all the holders of bonobos cooperate—which they do—it remains a daunting task, requiring frequent genetic exchange. Many individual apes owned by one institution are therefore on loan to another at any given time. Two programs coordinate breeding efforts: the Species Survival Plan (SSP) in North America and the European Endangered Species Programme (EESP).[1]

Most zoos work with small groups: currently, the average group size in captivity is only 5.6 individuals. The largest groups can be found at the Dierenpark Planckendael, in Mechelen, Belgium, and the Milwaukee County Zoo (Milwaukee, Wisconsin), but even these colonies will have to grow considerably before they can compare to the best-known chimpanzee colonies, some of which now include as many as thirty individuals. The small size of captive bonobo groups is, of course, partly owing to the relative lack of available apes, but also to ignorance of bonobos' social needs, combined with the desire of zoological societies to display this rarest of primates, which has resulted in a fragmentation of the population so as to satisfy as many zoos as possible. In 1993, however, the breeding programs adopted a policy of chiefly exchanging maturing females, leaving sons with mothers whenever possible, and building larger groups. It is to be hoped that this policy will result in more natural groupings that do justice to the species' complex sociality.[2]

One may be lucky enough to see interesting bonobo behavior during a brief visit to the zoo, but it generally requires quite a bit more effort to see the sort of social and sexual interactions described in this book. Ideally, one should visit a nearby zoo that hosts bonobos on a regular basis and learn to distinguish the apes' faces. Some zoos have a portrait gallery that will be of help in naming the individuals, or one may ask an animal caretaker for their names. There is also nothing wrong with assigning them names of one's own choosing. Individual identification is the first step towards getting to know bonobos and developing a feeling for their social relationships. Some individuals are close friends—that is, they frequently groom each other and support each other in fights. Others cannot stand each other: they stay away from each other and have regular skirmishes. All of this is discovered only gradually, because when observers say that a behavior occurs "frequently" or "regularly," they sometimes mean once or twice a day. A rare behavior may occur only once a week or month, and to see it will take even more patience.

The international smuggling of wild primates, once a major problem, has virtually come to a halt. Although ape infants are no longer caught for their export value, they are often removed from their mother after she has been shot for meat. Orphans taken as pets usually have a short, miserable life, but these two are well cared for in an ape orphanage in Kinshasa.

The joy of watching bonobos is tremendously increased when one sees a juvenile growing up and changing his or her behavior and social relationships. The longer one follows a group, the more its close-knit social life will come to seem like a soap opera, with happy and hilarious but also sad and disturbing moments. In the end, what will perhaps make the deepest impression on any amateur ape watcher is the incredible individual diversity among bonobos. Like people, they differ widely from one another in intelligence, temperament, and behavior. Once one recognizes them as individuals, one will quickly begin to see them as the different personalities that they are.

BIAS
EKON
MEZO
NJUNA
ZENTO

THE STRUGGLE FOR SURVIVAL

People mark their territories: along a trail in the forest around Wamba, local villagers have carved graffiti into a tree. Trees are part of the bonobo's everyday environment, representing safety and food. How much longer will Zaire's rain forest remain relatively untouched by humans?

A man from a village near Wamba prepares for a hunt by stretching his bow (above). He is after monkeys, duikers, and other forest animals but says he abides by the local taboo against killing bonobos. Even if they are not the target, however, bonobos run serious risks from these hunts. This female (opposite) lost an entire hand, most likely in a snare. A great many bonobos suffer such deformities.

Bonobos cross a road while schoolchildren watch in the background. Even in the heart of the bonobo's range, people now outnumber the apes. Coexistence is possible, however, as Wamba's experience shows. We can only hope that the local tradition of respect for the species will be passed on to future generations.

NOTES

CHAPTER 1. THE LAST APE

1. Cartmill 1993, 14.

2. Yerkes 1925, 246.

3. It is generally assumed that there is zero overlap between bonobo and chimpanzee populations owing to the river barrier (apes do not swim). Genetic analysis may help establish whether hybridization along the river took place in the past during periods when the river was perhaps not as wide.

4. Yerkes 1925, 244.

5. The fact that all the bonobos died of fear that night, whereas none of the zoo's many chimpanzees suffered a similar fate, attests to the extraordinary sensitivity of the former species (Tratz and Heck 1954).

6. Tratz and Heck 1954, 99.

7. Morris 1967, for example, claims that the "naked ape" (his label for our species) is the sexiest primate alive, the only one to know female orgasm, and that its males are endowed with the largest penis. Although I am unaware of any detailed measurements of bonobo penises, anyone seeing a male of this species with a full erection will doubt Morris's claim. The circumference of the bonobo's penis is less than that of the average man's, but the penis seems considerably longer, both absolutely and especially relative to the bonobo's smaller body. As for female orgasm and general "sexiness," readers are referred to chapter 4.

8. The birth weight of bonobos is only three-quarters that of chimpanzees. Thompson-Handler (1990) calculated a mean (± standard deviation) weight of 1,381 ± 199 g for eight captive-born bonobos, whereas chimpanzee newborns weigh around 1,800 g. See Kuroda 1989 for preliminary data on the bonobo's relatively slow growth rate.

9. Campbell 1980, 7.

CHAPTER 2. TWO KINDS OF CHIMPANZEE

1. Coolidge 1933, 56.

2. Greater locomotor efficiency for a bipedal gait, as proposed by Rodman and McHenry 1980, has been critically discussed by Streudel 1994. The savanna story about bipedality with which all of us grew up has become controversial as a direct result of newly discovered hominid fossils. See Shreeve 1996 for a popular review.

3. For information about Little Foot and the role of the forest habitat in human evolution, see Susman et al. 1984, Boesch and Boesch 1994, and Clarke and Tobias 1995.

4. Susman 1984, 390.

5. The greatest cytogenetic difference between humans and apes is that we have only 46 chromosomes, whereas apes have 48. In the course of evolution, humans fused two ancestral chromosomes into one. Comparison of karyotypes indicates that (a) apes have evolved more from the common ancestor than is generally realized, and (b) chromosomally, the bonobo is the most specialized African ape. On these grounds, Stanyon et al. 1986 questions the idea that *Pan paniscus* is the best extant model of the common ancestor.

6. In some scientific circles there is a tendency to compare apes with human children. Apes are often thought of as cute and playful creatures, constantly getting into mischief. This may be true of young apes, but it is quite mistaken with regard to adults. The neoteny argument follows exactly the opposite logic: it says that adult humans have more in common with ape children than with adult apes (de Waal 1989, 249–52). The idea of retention of juvenile traits in the course of evolution goes back to Louis Bolk's (1926) claim that *Homo sapiens* looks very much like a primate fetus that has reached sexual maturity. Neoteny is insightfully discussed in Gould 1977. See Shea 1983 and Blount 1990 for application of the neoteny hypothesis to bonobo evolution.

7. The masked or pale-faced chimpanzee (*Pan troglodytes verus*) lives in West Africa, the black-faced chimpanzee (*P. t. troglodytes*) in Central Africa, and the smaller long-haired chimpanzee (*P. t. schweinfurthii*) in East Africa. Chimpanzee DNA was analyzed by Morin et al. (1994), and bonobo DNA by Gerloff et al. (1995).

8. The results were published as a detailed, quantitative behavioral description, includ-

ing sound spectrograms, in the journal of *Behaviour* (de Waal 1988). A 54-minute videocassette, *Social Ethogram of the San Diego Bonobos,* illustrates the behavior patterns. A 6-minute audiocassette, *Vocal Repertoire of the Bonobo Compared with That of the Chimpanzee,* illustrates the calls. (Copies of both tapes can be borrowed or bought via the Primate Library of the Wisconsin Regional Primate Research Center, 1223 Capitol Court, Madison, WI 53715-1299, USA [telephone: 608/263-3512].) Apart from comparisons with the chimpanzee ethograms of Goodall 1968 and van Hooff 1973, my ethogram compares with bonobo behavior previously described by Patterson 1979, Kano and co-workers (e.g., Kano 1980; Kuroda 1980; Mori 1984), and especially Jordan 1977. The analysis categorizes behavior patterns based on the situation in which they typically occur. Thus, a vocalization often preceding attack behavior is classified as a threat, and a gesture associated with grooming is interpreted as affiliative.

9. Bonobos are acrobats for whom hands and feet are interchangeable grasping organs. They will use a foot to pick up something, hold an object, kick each other, masturbate, or reach out for contact. Chimpanzees can do the same, but do it less often, as if greater functional differentiation between hands and feet has been achieved. This is already evident at an early age. When Vauclair and Bard (1983) compared object manipulations in three seven-month-old individuals of different species—a child, a chimpanzee, and a bonobo—they found that the human infant performed more complex manipulations and that the two ape infants did not differ from each other. However, the extremities used for manipulation varied greatly, with both the child and the chimpanzee rarely using their feet, whereas the bonobo used his feet more than 40 percent of the time.

10. Baring of the teeth in defense is undoubtedly the original function. In the course of evolution, this powerful visual signal began to communicate submission and low status. In some species, appeasing and friendly intentions became part of its meaning, as argued here for the bonobo. Although van Hooff's (1972) analysis of the evolution of laughter and smiling did not consider the bonobo, it provides food for thought for anyone interested in the expression of emotions in animals and humans, a topic first seriously considered by Darwin (1872).

11. Some of these San Diego stories are cited in Heublein 1977. General information about attribution and perspective-taking (sometimes known as a "theory of mind") in apes and young children can be found in Buttersworth et al. 1991 and Whiten 1991. De Waal 1996 discusses how these capacities may relate to sympathy and empathy. See also chapter 6.

12. Mirror self-recognition is a controversial field. Whether gorillas pass the mark test was for a long time debatable, and what the test exactly means in terms of "self-awareness" is as yet unresolved. For an overview of the current debate, including data in support of self-recognition in gorillas, see Parker et al. 1994. Preliminary mirror studies on bonobos have been conducted by Westergaard and Hyatt (1994) and Walraven et al. (1995).

13. In 1993, C. P. van Schaik and his colleagues encountered an orangutan population in Suaq Balimbing, on the island of Sumatra, in which individuals manufactured tools and adjusted them to their needs. For example, the apes broke off branches and stripped them clean of leaves to extract honey from the nests of stingless bees or to forage for insects. These were the very first such observations of wild orangutans. They confirm decades of knowledge about the tool use of the species in captivity and put the technology of orangutans on a par with that of chimpanzees. The authors conclude that the cognitive capacities for flexible tool-use must go back at least to the common ancestor of orangutans, the African apes, and hominids. See van Schaik et al. 1996.

14. McGrew 1992 classifies the tool technologies of chimpanzees throughout Africa in support of the claim that each community has its own material culture. Further, see Nishida 1987 and Wrangham et al. 1994.

15. In humans, the right hemisphere specializes in parallel mental processing, control of emotional responses, and processing of faces, whereas the left hemisphere specializes in analytical thinking and language. For a review of laterality research on nonhuman primates, see Hopkins and Morris 1993. Hopkins and de Waal 1995 compares data on 21 bonobos at the Yerkes Primate Center and the San Diego Zoo.

16. Tomasello et al. 1993 argues that Kanzi possesses abilities absent in apes with little exposure to people. Comparing (a) mother-reared apes, (b) apes, including Kanzi, who grew up

in close contact with human caretakers, and (c) human children, the investigators found mother-reared apes to be the poorest at imitation of an experimenter. The human-reared apes, on the other hand, performed as well as the children. The authors speak of "encultur-ated" apes, suggesting that intense exposure to the human cultural environment brings out cognitive capacities that remain dormant when apes interact only with their own kind. An alternative explanation presents itself, though. Perhaps apes only imitate the species involved in their rearing. In other words, mother-reared apes may not pay much attention to people, yet may very well be inclined to imitate conspecifics.

17. Savage-Rumbaugh and Lewin 1994, 174.

CHAPTER 3. IN THE HEART OF AFRICA

1. The pros and cons of food provisioning are an ongoing debate (Asquith 1989). The most successful long-term projects on wild chimpanzees, by Jane Goodall and Toshisada Nishida, both in Tanzania, involve provisioning of limited amounts of bananas or sugarcane (initial provisioning of large amounts gave rise to violence). More recently, scientists have taken the trouble of habituating chimpanzees without provisioning. It takes more time, but the efforts have paid off, a prime example being the project of Christophe Boesch in Taï National Park in the Ivory Coast. The bonobo research in Lomako Forest, too, has been set up without provisioning (see above, pp. 79–82, and Fruth 1995, 37–38, for description of the habituation technique). Although this site has lacked continuity in the past, it has the potential of becoming another nonprovisioning success story. This would be all the more important inasmuch as Lomako, albeit not a pristine site, is remote from human populations and less disturbed than Wamba (Thompson-Handler et al. 1995).

2. Apart from infant duikers (*Cephalophus spp.*), bonobos at Lomako also occasionally eat adult duikers (Hohmann and Fruth 1993). Weighing up to 10 kg, this forest antelope is a relatively large prey item, in the range of the colobus monkeys eaten by chimpanzees.

3. Observations of both wild chimpanzees (Sugiyama 1988; Boesch 1991) and captive chimpanzees (de Waal 1994) show female bonding to be a distinct potential in this species. According to Parish 1996a, female bonding in chimpanzees is constrained by ecological opportunity.

4. Frequent intersexual grooming is reported by Badrian and Badrian 1984b, 335–36; Kano 1992, 190; and Thompson-Handler 1990. Although some of this grooming probably involves mother-son combinations (Furuichi and Ihobe 1994), a high level of grooming remains after deduction of known kin dyads from Kano's data. In relation to male bonding, it is interesting to note that in the study of the Badrians, the longest grooming session, which lasted over two hours, took place between two adult males. Similarly, Kano's data indicate that males at Wamba groom each other for longer stretches of time than do females.

5. The total numbers of identified individuals (*N*) in the best-known wild bonobo communities are given below, along with ratios of adult males to adult females (M:F). See van Elsacker et al. 1995 for a review of community and party sizes in the bonobo.

Site	*Community*	*N*	*M:F*	*Source*
Wamba	E[a]	75	20:20	Kano 1992
Wamba	E1	32	7:9	Ihobe 1992
Wamba	E2	36	8:11	Ihobe 1992
Lomako	Hedons	44	8:14	Thompson-Handler 1990
Lomako	Rangers	26	5:8	Thompson-Handler 1990
Lomako	Eyengo[b]	34	6:12	Fruth 1995

[a] Wamba's group E consisted of northern and southern subgroups, which split permanently into E1 and E2 after group E reached its maximum size of *N* = 75 reported here. The split occurred around 1982; the E1 and E2 group sizes are for 1987.

[b] The "Rangers" in Lomako were renamed the "Eyengo" after a stream in their home range. The data from Thompson-Handler 1990 concern this community in 1984–86, whereas the Fruth 1995 data are for 1994.

6. Nonhuman predation on great apes was long thought insignificant until recent reports of leopard and lion predation on chimpanzees (e.g., Boesch 1991). The fact that chimpanzees and bonobos build nests every night to sleep off the ground suggests that nocturnal predation may be a serious problem.

7. Kano 1992, 193.

8. Kano 1992, 176, recorded 325 antagonistic encounters at the Wamba feeding site, of which only 3.4 percent were by one adult or adolescent female against another; the overwhelming majority involved males as both aggressors and recipients. Female fights "were much more severe than those between other dyads, although they did not occur as frequently," Furuichi notes (1989, 194). "These tendencies may reflect the competitive nature of relationships between unrelated females."

9. Kano 1992, 183–84, describes the reversal between Koguma and Ude, noting, "A strong mother . . . adds a complication. Her son achieves an unreasonably high position. This elevated rank may be a sham and temporary phenomenon or it may be maintained. Interestingly, young sons seem to discern the capability and effect of their mothers very well, because the sons of low-profile mothers do not make these bold challenges." This tactical sensitivity is illustrated by the struggle between Ten and Ibo recounted by Furuichi (1992a): it can hardly be accidental that Ten challenged Ibo when Kame had become physically weakened.

10. I first noticed female dominance during a follow-up visit to the San Diego Zoo in 1985. A male who during my earlier studies had dominated a single adult female was now housed with two females, the older of whom clearly had the upper hand. If food was thrown into the enclosure, this female would select the best items before the male went anywhere near it. One might argue that this does not prove the female's dominance. Perhaps the male was merely being tolerant and respectful. However, this female also occasionally chased the male; never the other way around. The same feeding priority, chasing, and avoidance pattern in any other combination of individuals (e.g., between males) would without hesitation be classified in terms of dominance. Unless future resarch indicates that under certain circumstances, male bonobos are capable of controlling females, I would conclude that some of the older females in both captive and wild communities show all the signs of firmly outranking all the males. It is to be hoped that future research will clarify whether this female dominance is based on female alliances, male inhibitions, or seniority (female bonobos probably grow older than males, as also reported for chimpanzees by Dyke et al. 1995).

11. Kano 1992, 188.

12. Furuichi 1992a.

13. Four kinds of food transfer during interactions over plant food were distinguished among chimpanzees at the Yerkes Primate Center and bonobos at the San Diego Zoo. Definitions, and data showing that tolerant food transfers (i.e., relaxed taking and co-feeding) were more common in chimpanzees, are given in the table below. Although a breakdown shows the differences to be fairly general, it remains possible that bonobos are equally or more tolerant in a few specific categories of relationships, such as those among older females.

	Chimpanzees	*Bonobos*
Number of transfers	2,377	598
Forced claim or theft[a]	9.5%	44.5%
Relaxed taking[b]	37.1%	15.7%
Co-feeding[c]	35.9%	17.6%
Collecting nearby[d]	17.6%	22.2%

[a] An ape supplants another at a food source, grabs food by force, or snatches a piece and runs.

[b] An ape, in full view of the possessor, removes food from his or her hands in a relaxed or playful manner without threat signals or use of force.

[c] An ape joins the possessor to feed peacefully from the same source, which both may hold.

[d] An ape waits for dropped pieces and scraps, which are collected from within arm's reach of the possessor.

14. In Wamba, three bonobos lack both testicles. Although various explanations may apply, we cannot exclude the possibility of vicious attacks similar to those reported for chimpanzees (de Waal 1986; Goodall 1992). Kano documents an astonishing array of physical abnormalities in bonobos, most of which can be attributed to poachers' snares, poisonous snakes, and the like. But he also notes: "It is certain that males are more apt than females to sustain injuries. Males are major participants in conspecific aggression, and consequently tend more to engage in violent or acrobatic locomotion. As a result, they face an increased danger of meeting with accidents" (Kano 1984, 5).

15. Parish (in press) compiled records of injuries due to fighting in zoo colonies of bonobos. She found that all the wounds were caused when females, often collectively, attacked males. See also the interview with Parish on pp. 113–15.

16. Wrangham 1993, 71.

17. Long before most of us began to speculate about bonobo evolution, Kortlandt (1972, 15) made the same assumption: "The pygmy chimpanzee appears to be a fairly recent secondary adaptation to the periodically flooded swamp forests between the Congo, Lualaba, and Kasai Rivers, bordered in the south by the Katanga depressions with their lakes and swamps. The secondary nature of the adaptations of this species is suggested by its 'gibbonized' build, its sensitivity to food, and the presence of typically non-arboreal elements in its behaviour (e.g. knuckle-walking in the terrestrial way)."

18. My reference to the "motherland" is because the more pacific intercommunity relations of bonobos may have come about owing to female dominance. It is probably always in the interest of males to prevent group females from copulating with extragroup males. This restriction is not in the females' interest, however, as it limits mate choice. Once females achieved the upper hand, males may have lost control over this critical issue. Once copulations between males and females of different communities occur on a regular basis, this may reduce male competition over territories and the females contained therein. First, some of their competitors—the "enemy" males in neighboring territories—might well be their brothers, fathers, and sons. Second, males do not need risky fights to gain access to neighboring females if there are opportunities to fertilize them during intercommunity mingling. In short, sexual relations between groups may have removed some of the evolutionary advantages that males gain from intergroup warfare.

CHAPTER 4. APES FROM VENUS

1. Hockett and Ascher 1968, 34.
2. Wescott 1968, 92.
3. Jordan 1977, 175.
4. The San Diego Zoo colony is exceptional in that ventro-ventral sexual positions are more common than ventro-dorsal ones. This was already the case in this colony a decade before my studies began (see Patterson 1979). The percentages of various sexual and erotic patterns in my records of 698 sociosexual interactions are as follows:

Behavior Pattern[a]	*Partner Combination*	
	Heterosexual[b]	*Other*[c]
Ventro-ventral mount or GG-rubbing	81%	52%
Ventro-dorsal mount	17%	25%
Back-to-back rubbing	0%	5%
Fellatio	0%	3%
Mouth/tongue kissing	1%	8%
Genital massage	2%	7%

[a] In addition, masturbation occurred 39 times.
[b] Pairs of adults and/or adolescents of the opposite sex.
[c] Same-sex combinations and contacts involving sexually immature apes.

5. Goldfoot et al. 1980.

6. Savage-Rumbaugh and Wilkerson 1978, 337.

7. Eighty-four percent of the world's cultures allow a man to marry multiple wives (Whyte, 1978). Very few men in these cultures, however, have the resources to support a large family. In reality, then, most families in the world contain one man and one woman.

8. Based on 16 live births of bonobos in captivity, Thompson-Handler (1990) found a median gestation length of 244 days, with a range of from 227 to 277 days. Interbirth intervals at Wamba were discussed at a symposium (unpublished) in 1995 by Furuichi, who warned that more data are needed before we conclude that the shortness of the average bonobo interval (i.e., 4.5 years) distinguishes bonobos from chimpanzees. Data on wild chimpanzees mostly stem from relatively open habitats. The interbirth intervals of forest-living chimpanzees may be more like those of bonobos because of the similarity in environment and food availability.

9. Data on genital anatomy, menstrual cycle, and sexual activity stem from Savage and Bakeman 1978, Dahl 1986, Dahl et al. 1991, de Waal 1987, Furuichi 1987, Furuichi 1992b, Blount 1990, and Wrangham 1993.

10. Genital swellings probably evolved after the human-ape split, and only in the *Pan* lineage; like us, gorillas and orangutans have no or only very small genital swellings (see Hrdy and Whitten 1987 for a review of sexual advertisement in female primates). If this is true, we do not need to explain the "loss" of sexual swelling in our species, since we had none to begin with. It has also been suggested, however, that our ancestors did start out with swellings, which were gradually replaced by permanent buttocks (Szalay and Costello 1991), and that our species' breasts and fleshy lips mimic, respectively, buttocks and labia (Morris 1977). The relocation of signals from the back to the front may be related to face-to-face intercourse. It is interesting, therefore, that bonobos have pinkish lips too, as well as breasts that appear more protuberant than those of other apes.

11. Fisher 1983, 220.

12. Small 1993, 198.

13. Translated by Suehisa Kuroda from his own Japanese original, Kuroda 1982.

14. Kano 1992, 169.

15. The life stages of female wild bonobos, with ages estimated from knowledge of chimpanzees and of captive bonobos, are given in the table below. See Wrangham 1993 and Thompson-Handler 1990 for comparisons of bonobo and chimpanzee life histories.

Life Stage	*Age (years)*	*Source*
Nursing period	0–5	Kuroda, 1989
First genital swelling (onset of adolescence)	7	Kano, 1989
Begins to wander between groups	8	Kano, 1992
Settles into new group	9–13	Furuichi, 1989
Menarche and first full-sized swelling	10	(estimate)
Growth-cessation (adult size reached)[a]	14–16	Kuroda, 1989
First offspring	13–15	Kuroda, 1989
Cessation of ovulation (menopause)	40	(estimate)
Number of offspring possible in lifetime: 5 to 6		
Longevity	50–55	(estimate)

[a] Parish (in press) provides the most complete data set on weight development in captive bonobos.

16. Parish (1996b) found that captive female bonobos who were not transferred out of their natal groups during adolescence began to reproduce several years later than females who had been moved to another group.

17. Exceptions are a male bonobo's relations with paternal sisters and his own daughters. Since no early familiarity is established with these individuals, no sexual inhibitions develop. Hence the need for female migration. Many other primates solve the same problem through male migration (Pusey and Packer 1987).

18. Hashimoto and Furuichi 1994, 159.

19. The hard time experienced by young males may not be limited to sex. Fruth (1995, 126–28) documents two separate cases of pubertal male bonobos at Lomako who were denied access to food. They were systematically chased out of fruit trees, displaced from food sources, and had to content themselves with leftovers. Only when traveling alone with their mothers could they eat without competing. One of these males disappeared during the study. The investigator notes that young females may be able to overcome this problem through alliances with older females that allow them access to feeding trees.

20. Sugiyama 1967, 233. Sugiyama was the first to speculate about the origin of infanticide.

21. Estimates of the contribution of infanticide to infant mortality are derived from a recent review by E. H. M. Sterck, D. Watts, and C. P. van Schaik (Sterck 1995, 121). On the ongoing debate about infanticide in nonhuman primates, see *Evolutionary Anthropology* 2, 2 (1995).

22. Kano 1992, 208.

23. Kano described the twig-dropping at a 1994 symposium. Despite the positive relation between male rank and mating success, only 5.2 percent of mature copulations at Wamba's feeding site are aggressively interrupted (Kano 1996).

24. Wrangham 1993 and Parish 1996b. Kano first publicly speculated about countermeasures against infanticide in 1995 at an unpublished symposium.

CHAPTER 5. BONOBOS AND US

1. Haeckel quoted by Susman 1987, 85. Malinowski 1929. Diamond 1990, 441. The latter cites accounts of how the Hawaiians worshipped the genitals in song and dance and pampered these body parts in their children. Breast milk was squirted into an infant's vagina and the labia molded together so that they would not separate. A little girl's clitoris was stretched and lengthened through oral stimulation. The penises of little boys received similar treatment so as to enhance their beauty and prepare them for sexual enjoyment later in life.

Although some anthropologists, relying on informants and early explorers rather than on firsthand observation, have romanticized the subject (see, e.g., the criticism of Margaret Mead in Freeman 1983), unrestrained sexual hedonism is nonetheless unlikely in any human culture. A stable family organization is simply unthinkable without moral constraints on the expression of sexuality. Thus, even the—in Western eyes—sexually most liberal cultures are not free of jealousy and violence in response to extramarital affairs. Universally, intercourse tends to take place in private (Friedl 1994), and the genital region tends to be hidden from view so as to prevent unintentional arousal of members of the opposite sex. That even the early Hawaiians knew chastity is evident from their word for loincloth, *malo,* which most likely derives from the Malayan word for shame, *malu* (Wulf Schiefenhövel, personal communication).

2. Reynolds 1967b, 34–35.

3. Baboons are another primate species that has successfully conquered the same habitat. Some baboon species have one-male units in which a male continuously associates with several females (Kummer 1968), whereas others establish heterosexual "friendships" that protect females and young (Smuts 1985). It has been speculated that chimpanzeelike apes also tried to enter the plains. Because of continued reliance on the forest and perhaps less developed cooperation between the sexes, they may have been pushed back into their ancestral habitat by hominids making the same move more completely. If chimpanzees do indeed descend from apes that were on their way to but did not succeed in developing savanna adaptations, they could be said to be a product of dehumanization (Kortlandt and van Zon 1969).

4. Considering the primate order as a whole, both the *Hylobatidae* (gibbons and siamangs) and *Callitrichidae* (marmosets and tamarins) evolved mating systems that, as in our species, involves heterosexual bonding. One major difference remains, however: whereas human families cooperate within larger communities, the families of these primates are separate, territorial units.

5. Van Schaik and Dunbar (1990) speculate that guarding against infanticide may underlie monogamy in the gibbons and other primates. They depict infanticide as a hidden force in evolution, the effects of which remain unobserved unless counterstrategies fail. According to these authors, the human pair bond may also have started out as an anti-infanticidal measure.

Smuts (1992) proposes a related scenario in which the human pair bond initially served to defend females against rape, sexual harassment, and other kinds of male violence. If true, the sex-for-food hypothesis of Lovejoy (1981) and Fisher (1983) is best looked at as a secondary development. That is, the exchange of resources and services between mates builds upon the initial protective arrangement.

6. In wild chimpanzees, the females use tools both more frequently and more skillfully than do the males (McGrew 1979; Boesch and Boesch 1984).

7. The term *homosexual* is employed here strictly to denote sexual contact between members of the same sex. Sexual orientation or identity are not implied. In contrast to people, bonobos are, so far as we know, never exclusively oriented to the members of one sex or another. This means that none of these apes is homosexual or heterosexual in the usual sense: bonobos are literally pansexual.

8. The problem with a line of reasoning that considers male interests on the whole is that natural selection works with *relative* reproductive success. From each male's perspective, elimination of infanticide is worthwhile only if it enhances his own reproduction compared with that of other males.

9. Female dominance remains the most puzzling feature of bonobo society. Did female cooperation not suffice as protection against infanticide? Was dominating the males a necessary step on the road to complete protection, or was it to achieve other ends? It is easy to see the advantage for females to being able to monopolize food, but does not this advantage apply to a host of other species that never evolved female dominance? In male-dominated primates, access to food by females can be secured through other mechanisms, such as a high level of male tolerance to females. In short, it remains unclear why bonobos moved from female alliances, as known in many species, to actual female dominance.

One of the issues that needs closer investigation is to what degree female feeding priority is aggressively enforced. Since Yerkes 1941, we have known that a female chimpanzee on her own often manages to claim priority when genitally swollen: the male just steps back and lets her have the food. Could it be that bonobo males do the same? Since the females of their species are swollen much of the time, the advantage that females enjoy when sexually attractive may have been turned into a permanent condition.

CHAPTER 6. SENSITIVITY

1. Wassenaar Zoo in the Netherlands closed a couple of years later. Its bonobos went to the Milwaukee County Zoo in the United States.

2. It is tempting to speculate that bonobos are flooded with oxytocin. This mammalian hormone promotes the mother-offspring bond. Oxytocin is released during birth and in nursing females; its original function is to trigger muscle contractions necessary for parturition and lactation. When it reaches the brain, however, the hormone seems to serve a broader function. In rodents, oxytocin has been found to stimulate affectionate behavior and, in turn, to be released as a result of such behavior (Insel 1992). If oxytocin indeed circulates in large quantities in the blood of bonobos—a testable hypothesis—this might explain their high level of social and sexual bonding. That possibility is not, however, an alternative to the evolutionary scenarios discussed in chapter 5. If bonobos have physiological mechanisms that make them seek intensive contact, and that perhaps provide them with satisfaction from such contact, the question of how these mechanisms evolved still remains. It is generally assumed that natural selection "attaches" rewarding experiences to particular kinds of behavior so as to urge animals to do what is best for them.

3. This scene between Kanzi and Tamuli is shown in a television documentary entitled "Bonobo People," produced by NHK of Japan.

4. Yerkes 1925, 246.

EPILOGUE. BONOBOS TODAY AND TOMORROW

1. Few captive bonobos in the world fall outside the jurisdiction of the international breeding program of the SSP and EESP, which every two years issues the *International Stud-*

book of the Bonobo from which the data below were taken. As of January 1, 1996, there were the following bonobo holdings at zoos and research institutions:

Institution	*Country*	*Males*	*Females*	*Total*
Planckendael (Antwerp/Mechelen)	Belgium	5	4	9
Antwerp Zoo	Belgium	2	0	2
Berlin Zoo	Germany	2	2	4
Cincinnati Zoo	United States	3	3	6
Columbus Zoo	United States	4	2	6
Fort Worth Zoo	United States	3	0	3
Frankfurt Zoo	Germany	1	7	8
Cologne Zoo	Germany	2	3	5
Language Research Center (Atlanta)	United States	2	4	6
Leipzig Zoo	Germany	2	1	3
Milwaukee Zoo	United States	5	6	11
Morelia Zoo	Mexico	2	1	3
San Diego Wild Animal Park	United States	2	4	6
San Diego Zoo	United States	3	6	9
Stuttgart Zoo	Germany	4	4	8
Twycross Zoo	Great Britain	3	2	5
Wuppertal Zoo	Germany	4	1	5
Yerkes Primate Research Center (Atlanta)	United States	2	3	5
World Total		51	53	104

2. Copies of the *International Studbook of the Bonobo,* which contains detailed information on the bonobo collections of the world's zoological societies, are available from Bruno van Puijenbroeck, Royal Zoological Society of Antwerp, Koningin Astridplein 26, Antwerp, Belgium.

BIBLIOGRAPHY

Asquith, P. 1989. Provisioning and the study of free-ranging primates: History, effects, and prospects. *Yearbook of Physical Anthropology* 32: 129–58.

van den Audenaerde, D. F. E. T. 1984. The Tervuren Museum and the pygmy chimpanzee. In *The Pygmy Chimpanzee*, ed. R. L. Susman, 3–11. New York: Plenum Press.

Badrian, A., and N. Badrian. 1977. Pygmy chimpanzees. *Oryx* 13: 463–68.

———. 1984a. The Bonobo branch of the family tree. *Animal Kingdom* 87, 4: 39–45.

———. 1984b. Social organization of *Pan paniscus* in Lomako Forest, Zaire. In *The Pygmy Chimpanzee*, ed. R. L. Susman, 325–46. New York: Plenum Press.

Blount, B. G. 1990. Issues in bonobo (*Pan paniscus*) sexual behavior. *American Anthropologist* 92: 702–14.

Boesch, C. 1991. The effects of leopard predation on grouping patterns in forest chimpanzees. *Behaviour* 117: 220–42.

Boesch, C., and H. Boesch. 1984. Sex differences in the use of natural hammers by wild chimpanzees: A preliminary report. *Journal of Human Evolution* 13: 415–585.

Boesch, H., and C. Boesch. 1994. Hominization in the rainforest: The chimpanzee's piece of the puzzle. *Evolutionary Anthropology* 3, 1: 9–16.

Bolk, L. 1926. *Das Problem der Menschwerdung*. Jena: Gustav Fischer.

van Bree, P. J. H. 1963. On a specimen of *Pan paniscus* Schwarz, 1929, which lived in the Amsterdam Zoo from 1911 till 1916. *Zoologische Garten* 27: 292–95.

Buttersworth, G. E., P. L. Harris, A. M. Leslie, and H. M. Wellman. 1991. *Perspectives on the Child's Theory of Mind*. Oxford: Oxford University Press.

Campbell, S. 1980. Kakowet. *Zoonooz*, December: 7–11.

Cartmill, M. 1993. *A View to a Death in the Morning: Hunting and Nature through History*. Cambridge, Mass.: Harvard University Press.

Clarke, R. J., and P. V. Tobias. 1995. Sterkfontein member 2 foot bones of the oldest South African hominid. *Science* 269: 521–24.

Coolidge, H. J. 1933. *Pan paniscus:* Pygmy chimpanzee from south of the Congo River. *American Journal of Physical Anthropology* 18: 1–57.

———. 1984. Historical remarks bearing on the discovery of *Pan paniscus*. In *The Pygmy Chimpanzee*, ed. R. L. Susman, ix–xiii. New York: Plenum Press.

Dahl, J. F. 1986. Cyclic perineal swelling during the intermenstrual intervals of captive female pygmy chimpanzees (*Pan paniscus*). *Journal of Human Evolution* 15: 369–85.

Dahl, J.F., R. D. Nadler, and D. C. Collins. 1991. Monitoring the ovarian cycles of *Pan troglodytes* and *P. paniscus:* A comparative approach. *American Journal of Primatology* 24: 195–209.

Darwin, C. 1965 [1872]. *The Expression of the Emotions in Man and Animals*. Chicago: University of Chicago Press.

Diamond, M. 1990. Selected cross-generational sexual behavior in traditional Hawai'i: A sexological ethnography. In *Pedophilia: Biosocial Dimensions*, ed. J. R. Feierman, 378–93. New York: Springer.

Dyke, B., T. B. Gage, P. L. Alford, B. Swenson, and S. Williams-Blangero. 1995. Model life table for captive chimpanzees. *American Journal of Primatology* 37: 25–37.

van Elsacker, L., H. Vervaeke, and R. F. Verheyen. 1995. A review of terminology on aggregation patterns in bonobos (*Pan paniscus*). *International Journal of Primatology* 16: 37–52.

Fisher, H. 1983. *The Sex Contract: The Evolution of Human Behavior*. New York: Quill.

Freeman, D. 1983. *Margaret Mead and Samoa*. Cambridge, Mass.: Harvard University Press.

Friedl, E. 1994. Sex the Invisible. *American Anthropologist* 96: 833–44.

Fruth, B. 1995. *Nests and Nest Groups in Wild Bonobos (Pan paniscus): Ecological and Behavioral Correlates*. Aachen: Shaker.

Fruth, B., and Hohmann, G. 1994. Comparative analyses of nest building behavior in bonobos and chimpanzees. In *Chimpanzee Cultures*, ed. R. W. Wrangham, W. C. McGrew, F. B. M. de Waal, and P. Heltne, 109–28. Cambridge, Mass.: Harvard University Press.

Furuichi, T. 1987. Sexual swellings, receptivity, and grouping of wild pygmy chimpanzee females at Wamba, Zaire. *Primates* 28: 309–18.

———. 1989. Social interactions and the life history of female *Pan paniscus* in Wamba, Zaire. *International Journal of Primatology* 10: 173–97.

———. 1992a. Dominance status of wild bonobos (*Pan paniscus*) at Wamba, Zaire. Paper given at the 24th Congress of the International Primatological Society, Strasbourg, France.

———. 1992b. The prolonged estrus of females and factors influencing mating in a wild group of bonobos (*Pan paniscus*) in Wamba, Zaire. In *Topics in Primatology*, vol. 2: *Behavior, Ecology, and Conservation*, ed. N. Itoigawa, Y. Sugiyama, G. P. Sackett, and R. K. R. Thompson, 179–90. Tokyo: University of Tokyo Press.

Furuichi, T., and H. Ihobe. 1994. Variation in male relationships in bonobos and chimpanzees. *Behaviour* 130: 211–28.

Gallup, G. 1982. Self-awareness and the emergence of mind in primates. *American Journal of Primatology* 2: 237–48.

Gerloff, U., C. Schlötterer, K. Rassmann, I. Rambold, G. Hohmann, B. Fruth, and D. Tautz. 1995. Amplification of hypervariable simple sequence repeats (microsatellites) from excremental DNA of wild living bonobos (*Pan paniscus*). *Molecular Ecology* 4: 515–18.

Goldfoot, D. A., H. Westerborg-van Loon, W. Groeneveld, and A. K. Slob. 1980. Behavioral and physiological evidence of sexual climax in the female stump-tailed macaque (*Macaca arctoides*). *Science* 208: 1477–79.

Goodall, J. [van Lawick-]. 1968. The behaviour of free-living chimpanzees in the Gombe Stream Reserve. *Animal Behaviour Monographs* 1: 161–311.

———. 1992. Unusual violence in the overthrow of an alpha male chimpanzee at Gombe. In *Topics in Primatology*, vol. 1: *Human Origins*, ed. T. Nishida, W. C. McGrew, P. Marler, M. Pickford, and F. B. M. de Waal, 131–42. Tokyo: University of Tokyo Press.

Gould, S. J. 1977. *Ontogeny and Phylogeny*. Cambridge, Mass.: Harvard University Press, Belknap Press.

Hashimoto, C., and T. Furuichi. 1994. Social role and development of noncopulatory sexual behavior of wild bonobos. In *Chimpanzee Cultures*, ed. R. W. Wrangham, W. C. McGrew, F. B. M. de Waal, and P. Heltne, 155–68. Cambridge, Mass.: Harvard University Press.

Heublein, E. 1977. Kakowet's family. *Zoonooz*, October: 5–10.

Hockett, C. F., and R. Ascher. 1968 [1964]. The human revolution. In *Culture: Man's Adaptive Dimension*, ed. A. Montagu, 20–101. Oxford: Oxford University Press.

Hohmann, G., and B. Fruth. 1993. Field observations on meat sharing among bonobos (*Pan paniscus*). *Folia primatologica* 60: 225–29.

———. In press. Food sharing and status in provisioned bonobos (*Pan paniscus*): Preliminary results. In *Food and the Status Quest*, ed. P. Wiessner and W. Schiefenhövel. Oxford: Berghahn Publications.

van Hooff, J. A. R. A. M. 1972. A comparative approach to the phylogeny of laughter and smiling. In *Non-verbal Communication*, ed. R. A. Hinde, 209–41. Cambridge: Cambridge University Press.

———. 1973. A structural analysis of the social behaviour of a semi-captive group of chimpanzees. In *Expressive Movement and Non-verbal Communication*, ed. M. von Cranach and I. Vine, 75–162. London: Academic Press.

Hopkins, W. D., and F. B. M. de Waal. 1995. Behavioral laterality in captive bonobos (*Pan paniscus*): Replication and extension. *International Journal of Primatology* 16: 261–76.

Hopkins, W. D., and R. D. Morris. 1993. Handedness in great apes: A review of findings. *International Journal of Primatology* 14: 1–25.

Hrdy, S. B. 1979. Infanticide among animals: A review, classification, and examination of the implications for the reproductive strategies of females. *Ethology and Sociobiology* 1: 13–40.

Hrdy, S. B., and P. L. Whitten. 1987. Patterning of sexual activity. In *Primate Societies*, ed. B. Smuts et al., 370–84. Chicago: University of Chicago Press.

Idani, G. 1990. Relations between unit-groups of bonobos at Wamba: Encounters and temporary fusions. *African Study Monographs* 11: 153–86.

———. 1991. Social relationships between immigrant and resident bonobo (*Pan paniscus*) females at Wamba. *Folia primatologica* 57: 83–95.

Ihobe, H. 1992. Male-male relationships among wild bonobos (*Pan paniscus*) at Wamba, Republic of Zaire. *Primates* 33: 163–79.

Ingmanson, E. 1966. Tool-using behavior in wild *Pan paniscus*: Social and ecological considerations. In *Reaching into Thought: The Minds of the Great Apes*, ed. A. Russon, K. A. Bard, and S. T. Parker, 190–210. Cambridge: Cambridge University Press.

Insel, T. R. 1992. Oxytocin—A neuropeptide for affiliation: Evidence from behavioral, receptor autoradiographic, and comparative studies. *Psychoneuroendocrinology* 17: 3–35.

Jordan, C. 1977. Das Verhalten zoolebender Zwergschimpansen. Diss., Goethe University, Frankfurt.

———. 1982. Object manipulation and tool-use in captive pygmy chimpanzees (*Pan paniscus*). *Journal of Human Evolution* 11: 35–39.

Kano, T. 1980. Social behavior of wild pygmy chimpanzees (*Pan paniscus*) of Wamba: A preliminary report. *Journal of Human Evolution* 9: 243–60.

———. 1984. Observations of physical abnormalities among the wild bonobos (*Pan paniscus*) of Wamba, Zaire. *American Journal of Physical Anthropology* 63: 1–11.

———. 1989. The sexual behavior of pygmy chimpanzees. In *Understanding Chimpanzees*, ed. P. G. Heltne and L. A. Marquardt, 176–83. Cambridge, Mass.: Harvard University Press.

———. 1992. *The Last Ape: Pygmy Chimpanzee Behavior and Ecology.* Stanford, Calif.: Stanford University Press.

———. 1996. Male ranking order and copulation rate in a unit-group of bonobos at Wamba, Zaire. In *Great Ape Societies*, ed. W. C. McGrew, L. Marchant, and T. Nishida, 135–45. Cambridge: Cambridge University Press.

Kortlandt, A. 1972. *New Perspectives on Ape and Human Evolution.* Amsterdam: Stichting voor Psychobiologie, University of Amsterdam.

Kortlandt, A., and J. C. J. van Zon. 1969. The present state of research on the dehumanization hypothesis of African ape evolution. In *Proceedings of the 2nd Congress of the International Primatalogical Society, Atlanta (Ga.)*, 3: 10–13. Basel: Karger.

Kummer, H. 1968. *Social Organization of Hamadryas Baboons: A Field Study.* Chicago: University of Chicago Press.

Kuroda, S. 1979. Grouping of the pygmy chimpanzee. *Primates* 20: 161–83.

———. 1980. Social behavior of the pygmy chimpanzees. *Primates* 21: 181–97.

———. 1982. *The Unknown Ape: The Pygmy Chimpanzee* (in Japanese). Tokyo: Chikuma-Shobo.

———. 1984. Interaction over food among pygmy chimpanzees. In *The Pygmy Chimpanzee*, ed. R. L. Susman, 301–24. New York: Plenum Press.

———. 1989. Developmental retardation and behavioral characteristics in the pygmy chimpanzees. In *Understanding Chimpanzees*, ed. P. G. Heltne and L. A. Marquardt, 184–93. Cambridge, Mass.: Harvard University Press.

Lasswell, H. 1936. *Who Gets What, When and How.* New York: McGraw-Hill.

Learned, B. 1925. Voice and "language" of young chimpanzees. In *Chimpanzee Intelligence and Its Vocal Expressions*, ed. R. M. Yerkes and B. Learned, 57–157. Baltimore: Williams & Wilkins.

Lovejoy, C. O. 1981. The origin of man. *Science* 211: 341–50.

Malenky, R. K., and R. W. Wrangham. 1994. A quantitative comparison of terrestrial herbaceous food consumption by *Pan paniscus* in the Lomako Forest, Zaire, and *Pan troglodytes* in the Kibale Forest, Uganda. *American Journal of Primatology* 32: 1–12.

Malinowski, B. 1929. *The Sexual Life of Savages.* London: Lowe & Brydone.

McGrew, W. C. 1979. Evolutionary implications of sex-differences in chimpanzee predation and tool-use. In *The Great Apes*, ed. D. A. Hamburg and E. R. McCown, 440–63. Menlo Park, Calif.: Benjamin Cummings.

———. 1992. *Chimpanzee Material Culture.* Cambridge: Cambridge University Press.

Mori, A. 1984. An ethological study of pygmy chimpanzees in Wamba, Zaire: A comparison with chimpanzees. *Primates* 25: 255–78.

Morin, P. A., J. J. Moore, R. Chakraborty, L. Jin, J. Goodall, and D. S. Woodruff. 1994. Kin selection, social structure, gene flow, and the evolution of chimpanzees. *Science* 265: 1193–1201.

Morris, D. 1967. *The Naked Ape.* New York: Dell.

———. 1977. *Manwatching: A Field Guide to Human Behaviour.* London: Jonathan Cape.

Nishida, T. 1987. Local traditions and cultural transmission. In *Primate Societies,* ed. B. Smuts et al., 462–74. Chicago: University of Chicago Press.

Parish, A. R. 1993. Sex and food control in the "uncommon chimpanzee": How bonobo females overcome a phylogenetic legacy of male dominance. *Ethology and Sociobiology* 15: 157–79.

———. 1996a. Female relationships in bonobos (*Pan paniscus*): Evidence for bonding, cooperation, and female dominance in a male-philopatric species. *Human Nature* 7: 61–96.

———. 1996b. Timing of first reproduction in female bonobos (*Pan paniscus*). *American Journal of Primatology.*

———. In press. Sexual dimorphism, maturation, and female dominance in bonobos (*Pan paniscus*). *American Journal of Physical Anthropology.*

Parish, A. R., and F. B. M. de Waal. Under review. Social relationships in the bonobo (*Pan paniscus*) redefined: Evidence for female-bonding in a "non-female bonded" primate. *Behaviour.*

Parker, S. T., R. W. Mitchell, and M. L. Boccia, eds. 1994. *Self-Awareness in Animals and Humans: Developmental Perspectives.* Cambridge: Cambridge University Press.

Patterson, T. 1979. The behavior of a group of captive pygmy chimpanzees (*Pan paniscus*). *Primates* 20: 341–54.

Povinelli, D. J., S. T. Boysen, and K. E. Nelson. 1990. Inferences about guessing and knowing by chimpanzees (*Pan troglodytes*). *Journal of Comparative Psychology* 104: 203–10.

Pusey, A. E., and C. Packer. 1987. Dispersal and philopatry. In *Primate Societies,* ed. B. B. Smuts, D. L. Cheney, R. M. Seyfarth, R. W. Wrangham, and T. T. Struhsaker, 250–66. Chicago: University of Chicago Press.

Reynolds, V. 1967a. On the identity of the ape described by Tulp, 1641. *Folia primatologica* 5: 80–87.

———. 1967b. *The Apes.* New York: Dutton.

Rodman, P. S., and H. M. McHenry. 1980. Bioenergetics and the origin of hominid bipedalism. *American Journal of Physical Anthropology* 52: 103–6.

Sabater-Pi, J., M. Bermejo, G. Illera, and J. J. Vea. 1993. Behavior of bonobos (*Pan paniscus*) following their capture of monkeys in Zaire. *International Journal of Primatology* 14: 797–803.

Savage, S., and R. Bakeman. 1978. Sexual morphology and behavior in *Pan paniscus.* In *Proceedings of the 6th International Congress of Primatology,* 613–16. New York: Academic Press.

Savage-Rumbaugh, S., and R. Lewin. 1994. *Kanzi: The Ape at the Brink of the Human Mind.* New York: Wiley.

Savage-Rumbaugh, S., and B. Wilkerson. 1978. Socio-sexual behavior in *Pan paniscus* and *Pan troglodytes:* A comparative study. *Journal of Human Evolution* 7: 327–44.

van Schaik, C. P., and R. I. M. Dunbar. 1990. The evolution of monogamy in large primates: A new hypothesis and some crucial tests. *Behaviour* 115: 30–62.

van Schaik, C. P., E. A. Fox, and A. F. Sitompul. 1996. Manufacture and use of tools in wild Sumatran orangutans: Implications for human evolution. *Naturwissenschaften* 83: 186–88.

Schwarz, E. 1929. Das Vorkommen des Schimpansen auf den linken Kongo-Ufer. *Revue de zoologie et de botanique africaines* 16: 425–26.

Shea, B. T. 1983. Paedomorphosis and neotony in the pygmy chimpanzee. *Science* 222: 521–22.

Shreeve, J. 1996. Sunset on the Savanna. *Discover* 17, 7: 116–25.

Small, M. F. 1993. *Female Choices: Sexual Behavior of Female Primates.* Ithaca, N.Y.: Cornell University Press.

Smuts, B. B. 1985. *Sex and Friendship in Baboons.* New York: Aldine.

———. 1992. Male aggression against women. *Human Nature* 3: 1–44.

Sommer, V. 1994. Infanticide among the langurs of Jodhpur: Testing the sexual selection hypothesis with a long-term record. In *Infanticide and Parental Care,* ed. S. Parmigiani and F. S. vom Saal, 155–87. Chur, Switzerland: Harwood.

Stanyon, R., B. Chiarelli, K. Gottlieb, and W. H. Patton. 1986. The phylogenetic and taxo-

nomic status of *Pan paniscus:* A chromosomal perpective. *American Journal of Physical Anthropology* 69: 489–98.
Sterck, E. H. M. 1995. Females, foods and fights. Diss., University of Utrecht.
Streudel, K. 1994. Locomotor energetics and hominid evolution. *Evolutionary Anthropology* 3, 2: 42–48.
Sugiyama, Y. 1967. Social organization of Hanuman langurs. In *Social Communication among Primates,* ed. S. A. Altmann, 221–53. Chicago: University of Chicago Press.
———. 1988. Grooming interactions among adult chimpanzees at Bossou, Guinea, with special reference to social structure. *International Journal of Primatology* 9: 393–408.
Susman, R. L. 1984. The locomotor behavior of *Pan paniscus* in the Lomako Forest. In *The Pygmy Chimpanzee,* ed. R. L. Susman, 369–93. New York: Plenum.
———. 1987. Pygmy chimpanzees and common chimpanzees: Models for the behavioral ecology of the earliest hominids. In *The Evolution of Human Behavior: Primate Models,* ed. W. G. Kinzey, 72–86. Albany: State University of New York Press.
Susman, R. L., J. T. Stern, and W. L. Jungers. 1984. Arboreality and bipedality in the Hadar hominids. *Folia primatologica* 43: 113–56.
Suzuki, A. 1971. Carnivority and cannibalism observed among forest-living chimpanzees. *Journal of the Anthropological Society of Nippon* 79: 30–48.
Szalay, F. S., and R. K. Costello. 1991. Evolution of permanent estrus displays in hominids. *Journal of Human Evolution* 20: 439–64.
Thompson-Handler, N. 1990. The Pygmy Chimpanzee: Sociosexual Behavior, Reproductive Biology and Life History Patterns. Diss., Yale University.
Thompson-Handler, N., R. K. Malenky, and G. E. Reinartz. 1995. *Action Plan for* Pan paniscus: *Report on Free Ranging Populations and Proposals for Their Preservation.* Milwaukee: Zoological Society of Milwaukee County.
Tomasello, M., S. Savage-Rumbaugh, and A. C. Kruger. 1993. Imitative learning of actions on objects by children, chimpanzees, and enculturated chimpanzees. *Child Development* 64: 1688–1705.
Tratz, E. P., and H. Heck. 1954. Der afrikanische Anthropoide "Bonobo": Eine neue Menschenaffengattung. *Säugetierkundliche Mitteilungen* 2: 97–101.
Vauclair, J., and K. Bard. 1983. Development of manipulations with objects in ape and human infants. *Journal of Human Evolution* 12: 631–45.
de Waal, F. B. M. 1986. The brutal elimination of a rival among captive male chimpanzees. *Ethology and Sociobiology* 7: 237–51.
———. 1987. Tension regulation and nonreproductive functions of sex among captive bonobos (*Pan paniscus*). *National Geographic Research* 3: 318–335.
———. 1988. The communicative repertoire of captive bonobos (*Pan paniscus*), compared to that of chimpanzees. *Behaviour* 106: 183–251.
———. 1989 [1982]. *Chimpanzee Politics: Power and Sex among Apes.* Baltimore: Johns Hopkins University Press.
———. 1989. *Peacemaking among Primates.* Cambridge, Mass.: Harvard University Press.
———. 1992. Appeasement, celebration, and food sharing in the two *Pan* species. In *Topics in Primatology,* vol. 1: *Human Origins,* ed. T. Nishida, W. C. McGrew, P. Marler, M. Pickford, and F. B. M. de Waal, 37–50. Tokyo: University of Tokyo Press.
———. 1994. The chimpanzee's adaptive potential: A comparison of social life under captive and wild conditions. In *Chimpanzee Cultures,* ed. R. W. Wrangham, W. C. McGrew, F. B. M. de Waal, and P. Heltne, 243–60. Cambridge, Mass.: Harvard University Press.
———. 1995. Sex as an alternative to aggression in the bonobo. In *Sexual Nature, Sexual Culture,* ed. P. Abramson and S. Pinkerton, 37–56. Chicago: University of Chicago Press.
———. 1996. *Good Natured: The Origins of Right and Wrong in Humans and Other Animals.* Cambridge, Mass.: Harvard University Press.
Walraven, V., L. van Elsacker, and R. Verheyen. 1995. Reactions of a group of pygmy chimpanzees (*Pan paniscus*) to their mirror-images: Evidence of self-recognition. *Primates* 36: 145–50.

Wescott, R. W. 1968 [1963]. In *Culture: Man's Adaptive Dimension,* ed. A. Montagu, 91–93. Oxford: Oxford University Press.

Westergaard, G. C., and C. W. Hyatt. 1994. The responses of bonobos (*Pan paniscus*) to their mirror images: Evidence of self-recognition. *Human Evolution* 9: 273–79.

White, F. J., and C. A. Chapman. 1994. Contrasting chimpanzees and bonobos: Nearest neighbor distances and choices. *Folia primatologica* 63: 181–91.

White, F. J., and R. W. Wrangham. 1988. Feeding competition and patch size in the chimpanzee species *Pan paniscus* and *P. troglodytes. Behaviour* 105: 148–64.

Whiten, A. 1991. *Natural Theories of Mind: Evolution, Development and Simulation of Everyday Mindreading.* Oxford: Blackwell.

Whyte, M. K. 1978. Cross-cultural codes dealing with the relative status of women. *Ethnology* 17: 211–37.

Wrangham, R. W. 1986. Ecology and social relationships in two species of chimpanzee. In *Ecology and Social Evolution: Birds and Mammals,* ed. D. I. Rubenstein and R. W. Wrangham, 353–78. Princeton, N.J.: Princeton University Press.

———. 1993. The evolution of sexuality in chimpanzees and bonobos. *Human Nature* 4: 47–79.

Wrangham, R. W., W. C. McGrew, F. B. M. de Waal, and P. Heltne, eds. 1994. *Chimpanzee Cultures.* Cambridge, Mass.: Harvard University Press.

Yerkes, R. M. 1925. *Almost Human.* New York: Century.

———. 1941. Conjugal contrasts among chimpanzees. *Journal of Abnormal and Social Psychology* 36: 175–99.

Zihlman, A. L. 1984. Body build and tissue composition in *Pan paniscus* and *Pan troglodytes,* with comparisons to other hominoids. In *The Pygmy Chimpanzee,* ed. R. L. Susman, 179–200. New York: Plenum.

Zihlman, A. L., J. E. Cronin, D. L. Cramer, and V. M. Sarich. 1978. Pygmy chimpanzee as a possible prototype for the common ancestor of humans, chimpanzees, and gorillas. *Nature* 275: 744–46.

ACKNOWLEDGMENTS

This book would not have been possible without the enthusiastic cooperation of some of the most important international bonobo experts. Only a handful of scientists can call themselves that, and each one has contributed unique insights to the picture of bonobo society sketched in these pages. Some of these experts work in the field, others at zoos, and still others in laboratories, but all have been willing to share their thoughts and findings. Particular thanks are due to our six interviewees, most of whom also kindly read and commented on sections of the manuscript: Barbara Fruth, Gottfried Hohmann, Takayoshi Kano, Suehisa Kuroda, Amy Parish, and Sue Savage-Rumbaugh.

Further input was received from Marianne Holtkötter, Richard Malenky, Gay Reinartz, Wulf Schiefenhövel, Meredith Small, and Nancy Thompson-Handler. While we remain ultimately responsible for the book's contents, we thank all of the above colleagues for pointing out factual errors, providing additional information, and making us express certain thoughts more clearly.

Frans de Waal's bonobo research, conducted at the San Diego Zoo, was supported by the National Geographic Society and the Wisconsin Regional Primate Research Center in Madison. A generous fellowship from the Carl Friedrich von Siemens Stiftung enabled me to take a leave of absence in Munich for a reclusive existence devoted to writing, interrupted by strolls through the Englischer Garten. (The location is of symbolic relevance, for the first behavioral studies of bonobos took place in that very city, in the 1930s.) I thank Wulf and Grete Schiefenhövel, Marianne Oertl, Heinrich and Weibke Meier, and other Bavarian residents for their warm hospitality. My wife, Catherine Marin, provided loving company despite pressing duties elsewhere and read all first drafts. I am also grateful to German, Austrian, and Dutch audiences for questions and comments after lectures about bonobos; their feedback greatly helped shape the discussions in the book.

Frans Lanting's expedition to the bonobos of Wamba, Zaire, was supported by the National Geographic Society. For help in Africa, I am indebted to Takayoshi Kano, Takeshi Furuichi, Chie Hashimoto, Ellen Ingmanson, the late Father Piet at Yalisele Mission, Harry Goodall, Mike Chambers, Delfi Messinger, Mission Aviation Fellowship, Institut Zaïrois pour la Conservation de la Nature, Institut National de la Recherche Biomédicale, Mankoto ma Mbaelele, and Karl and Kathrine Ammann and Mzee. In the United States, I am grateful for help from Zoo Atlanta, Columbus Zoo, Cincinnati Zoo, San Diego Zoo, San Diego Wild Animal Park, the Language Research Center at Georgia State University, Adrienne Zihlman, Amy Parish, Geza Teleki, Bob Caputo, Isabel Stirling, and Nikon, Inc. In the Netherlands, I would like to thank Anton van Hooff at Burgers Zoo and the able doctors at Rotterdam Harbor Hospital who cured me of a serious case of cerebral malaria. For assistance at the National Geographic Society, I am grateful to former editors Bill Garrett and Bill Graves, Mary Smith, Bill Allen, Tom Kennedy, Kent Kobersteen, Al Royce, Neva Folk, Ann Judge, Jane Vessels, and the Photo Equipment Shop. Special thanks go to Sue Savage-Rumbaugh and Duane Rumbaugh, the staff of Minden Pictures, Julia Belanger, Kristine Asuncion, Michelle Reynolds, and the staff at Frans Lanting Photography. Christine Eckstrom's editorial judgment was invaluable with respect to bringing my own vision into focus. Last but not least, let me thank Kanzi, Lana, Loretta, Sen, Panbanisha, Akili, and all the other bonobos who helped blur the boundary between apes and humans before my eyes.

FRANS B.M. DE WAAL
Atlanta, Georgia

FRANS LANTING
Santa Cruz, California

PHOTOGRAPH LOCATIONS

Artis Zoo, Amsterdam, Netherlands: page 4
Burgers Dierenpark Arnhem, Netherlands: page 21
Central Zaire: pages 31, 56
Cincinnati Zoo: pages x–xi, 170
Columbus Zoo: page 146-left
Kahuzi Biega National Park, Zaire: page 19
Language Research Center, Georgia State University: pages 39, 45, 46
Orphanage of Delfi Messinger, Kinshasa, Zaire: pages 130, 144–45, 146-right, 177
Private care, Kenya: page 20
Sabah, Borneo: page 17
San Diego Zoo and San Diego Wild Animal Park: pages vi–vii, xvi, 51, 52, 54, 55, 98, 102, 111, 121, 124–25, 128–29, 131, 132, 147, 152, 155, 162–63, 164, 166, 167, 184, 212
Wamba, Zaire: pages ii, viii–ix, xii, 14–15, 22, 48, 49, 50, 53, 61, 69, 73, 77, 83, 84, 90–91, 92–93, 94, 95, 96, 97, 106, 126, 127, 141, 148, 149, 150–51, 165, 168, 169, 172, 175, 178–79, 180, 181, 182–83
Yerkes Primate Center, Atlanta: page 9
Zoo Atlanta: pages 16, 18

INDEX

HOW YOU CAN HELP

Further information about the conservation status of bonobos in the wild can be found in the "Action Plan for *Pan paniscus*" (Thompson-Handler et al. 1995), which can be obtained from Dr. Gay Reinartz, Bonobo SSP Coordinator, Zoological Society of Milwaukee County, 10005 W. Bluemound Road, Milwaukee, WI 53226, USA.

Readers interested in donating money towards bonobo conservation may contact the nearest zoological park that exhibits the species. Several zoological societies support field research or conservation projects. In addition, the following organizations specifically support bonobo conservation and protection in Zaire:

The Bonobo Protection and Conservation Fund
Laboratory of Human Evolution Studies,
(c/o Dr. Suehisa Kuroda),
Faculty of Science,
Kyoto University,
Sakyo, Kyoto, 606 JAPAN

Language Research Center
(c/o Dr. Sue Savage-Rumbaugh),
Georgia State University Foundation,
University Plaza,
Atlanta, GA 30303-3083, USA

Zoological Society of Kinshasa
Delfi Messinger,
c/o P. O. Box 1106,
Port Isabel, TX 78578, USA

Conservation Fund of the American Society of Primatologists (ASP)
Dr. Ramon Rhine, Chair,
Psychology Department,
University of California,
Riverside, CA 92521, USA

FOR MORE INFORMATION

On the World Wide Web, visit the following sites on bonobos and other primates:

http://www.primate.wisc.edu/pin/
(Primate Info Net of the Wisconsin Primate Center)
http://weber.u.washington.edu/wcalvin/bonobo.html
(some bonobo zoo exhibits)
http://jinrui.zool.kyoto-u.ac.jp/PAN/home.html
(Newsletter of the Japan Center for the Conservation and Care of Chimpanzees)
http://www.asp.org/
(home page of the American Society of Primatologists)

For information about Frans Lanting's work, contact Terra Editions, P. O. Box 409, Davenport, CA 95017, USA, or visit his World Wide Web site at http://www.lanting.com.

Designer: Barbara Jellow
Compositor: G&S Typesetters, Inc.
Text: 9.5/16 Cycles
Display: Trajan
Printer and Binder: Paramount Printing Co., Ltd.